Botanical Companions

American Land and Life

EDITED BY WAYNE FRANKLIN

# Botanical

## A Memoir of Plants

# Companions

and Place by Frieda E. Knobloch

FOREWORD BY WAYNE FRANKLIN

University of Iowa Press | Iowa City

University of Iowa Press, Iowa City 52242


Printed in the United States of America
Design by April Leidig-Higgins

http://www.uiowa.edu/uiowapress



The photographs on pages 13, 68–71, and 101 are by Michael Tierney.

A version of "Specimens" was previously published in the *Odense American Studies International Series* (OASIS), no. 59 (August 2003). A version of "Collecting" was previously published as "Collecting: The Work of Learning, Following Botanists" in *Prospects* 28 (2004): 221–44.

The University of Iowa Press is a member of Green Press Initiative and is committed to preserving natural resources.

Printed on acid-free paper

Library of Congress Cataloging-in-Publication Data
Knobloch, Frieda E.
Botanical companions: a memoir of plants and place / by Frieda E. Knobloch; foreword by Wayne Franklin.
p. cm.—(American land and life series)
Includes bibliographical references (p. ).
ISBN 0-87745-920-7 (cloth)
1. Nelson, Aven, 1859–1952. 2. Nelson, Ruth Ashton, 1896–. 3. Botanists—Wyoming—Biography. 4. Botany—Rocky Mountains Region. 5. Knobloch, Frieda E.—Travel—Rocky Mountains Region. I. Title. II. Series.
QK26.K555 2004
580'.92'2—dc22 2004049829
[B]

05 06 07 08 09 C 5 4 3 2 1

In Memoriam

Hazel Fargo (1901–1983)

Aven Nelson (1859–1952) and
Ruth Ashton Nelson (1896–1987)

Guides Close By and Far Away

# Contents

# Foreword

WAYNE FRANKLIN

In this fascinating study of the creation and preservation of botanical knowledge, Frieda Knobloch most of all asks us to consider how we image—and imagine—the order of the world. The sciences propose many answers to her question. Anyone who has taken introductory chemistry classes in high school or college is of course familiar with the periodic table of the elements, perhaps the most compelling graphic model of scientific learning. A marvel of compactness, it gives essential information about all of the currently enumerated elements (and some merely hypothesized ones, such as *ununseptium* or *ununoctium*) and displays their interrelations. As a humanist with an interest in science who sometimes teaches large American literature courses in a chemistry lab where a large periodic table hovers behind me as I teach poetry or fiction from the experiment bench, I must admit that I am envious of such devices. At times I even urge my students to contemplate how they would go about inventing literary or cultural counterparts for them.

A similar sense of humanistic envy compelled former colleagues of mine some years ago to produce a series of convenient literary handbooks called *Elements of Fiction*, *Elements of the Essay*, and so on. When they aggregated them into an Oxford University Press textbook called, of course, *Elements of Literature*, the metaphor tying the library desk to the laboratory bench became obvious. Part of the appeal of the project was the wish to organize humanistic subjects with what a literary scholar might regard as chemical precision. If only one could! But both terms in this envious relationship—literature and chemistry—are in fact more complex than they appear.

In the case of my own students, I hardly expect them to produce such a "table of literary elements." Indeed, the essentially heuristic exercise I assign them has value because it prompts students to think about the quite different models of learning that are appropriate to different fields. If reduced to graphic summary, humanistic models would, I think, be less overtly ordered. They might emphasize methods of connection rather than structures of relation. They would almost certainly be three- or perhaps even four-dimensional. And they would be more likely to take hints from excavational sciences such as archaeology or dendochronology than from chemistry. With all deference to my "elemental" colleagues, I don't believe a writer produces a new literary text by bringing a certain quantity of pastoral, say, into the neighborhood of a suitable batch of unbonded satire. Nor does one read by means of an electron microscope, seeking out the irreducible first particles of the work by means of their spectral traces in the reader's mind.

A reader perhaps comes closest to scientific method when she unfolds all the possible meanings of a given term or phrase as it is encountered on the page—with their sundry points of historical origin and their charming irrelevancies—and then proceeds to eliminate those which do not seem to be indicated in the current deployment of the word or phrase in question. The word is not, of course, just a sound given graphic presence on the page. It is the descendant of all prior uses, the source of all future ones. Through it rush energies from the dead generations to those not yet born. "Elements" doesn't quite capture the energies that flow through words and works.

Yet such speculations can cut both ways. When one reflects on that chemical model, the periodic table, one is at first impressed by its Platonic idealism. It seeks to represent the accumulated *Is*-ness of chemical learning. There is no room here for rejected elements and, except in rare instances such as those noted above, no room for the as-yet uncovered, either. What first strikes me as a humanist, a practitioner of one of the cultural sciences, is how little background the table appears to offer. From the chart on the wall, one knows virtually nothing about the complexly collaborative process by which this episteme has been constructed. One also misses completely the root metaphor of the table itself—the idea of recursive order that the British researcher John A. R. Newlands introduced with the first version of the chemical table in 1867. Newlands, noticing what he termed an "octave" effect in the elements, arranged the constituent elements in rows of seven, starting new lines to mark the repetitions. Two years later, the Russian chemist Dimitri Mendeleev refined the "octave" principle, positing that elements varied *periodically* according to their atomic weights. He introduced blank spaces into the chart because his theory predicted the existence of as-yet-unknown elements that in time were indeed discovered and duly placed in the blanks. What seems like a complex theoretic design to a sophomore literature major whose mind drifts away from *The Scarlet Letter* to the green lanthanoids and actinoids at the bottom of the periodic chart—just about where my head is in that class—is in part the result of a poetic thought, an analogy from musical practice and vibratory relations among sounds to the quite different world of "pure" matter.

To be sure, the Newlands-Mendeleev analogy did not hold beyond the first few rows, which is why the periodic table now has holes and outriggers and indeed does exist in a three-dimensional version aimed at displaying complex relations beyond those first suspected by such early proponents of the device. Moreover, the harder one pushes to divide humanistic protocols from scientific ones, the more one discovers that all learning proceeds by intuition and elaboration as well as by method. In a nicely interactive online version of the periodic chart produced by Web Elements, a viewer can learn the pertinent history of each currently recognized chemical element. By clicking on various menu items, one thus may learn of the role of Joseph Priestley and Carl Wilhelm Scheele in the discovery of oxygen in 1774. Suddenly the present structure of chemical learning is given vivid historical depth. To be sure, only rarely does science dwell on wholly outmoded models: not even Web Elements, keyed as it is to the current table, teaches one anything about Georg Ernst Stahl's pre-oxygen speculations about the imaginary element "phlogiston"! In this sense, perhaps, even the most elaborate versions of the periodic table repeat the key refrain of scientific knowledge: what is pertinent is what exists, not what various people at various times have wrongly ventured.

WHAT IS ESPECIALLY valuable about Frieda Knobloch's study of the botanical field work of Aven Nelson and Ruth Ashton Nelson in the Rocky Mountains from the late nineteenth to the mid twentieth century is, first, its insistence that learning always has a biographical context. Ideas do not hold themselves. They are held by minds—first by one or perhaps two, then, through various forms of dissemination, by others, perhaps many others. More intriguingly, Knobloch goes beyond her splendid but necessarily fragmentary biographical reconstruction of the work of these two botanists to dissect the dominant tropes of learning and remembering that too often are submerged beneath the surface of academic inquiry. Her discussion of field identification and collection practices and of the methods by which we physically model the world is so rich, so redolent of what I am tempted to call the spiritual practice of learning that I frankly know of nothing like it. After encountering Knobloch's deep, allusive revelations, one's view of the world does not easily resume its customary form.

Botanical Companions

Erysimum nivale
Leucocrinum montanum
Erigeron
Senecio integerrimus
Oxytropis campestris
Astralagus ceramicus
Bluebells
Yarrow
Pussytoes
Dandelions
Delphinium
2-flowered white star
Viola nuttallii
Mertensia ciliata
Oreoxis alpina
Sedum lanceolatum
Phlox multiflora
White aromatic umbel
Currant
Ranunculus glaberrimus
var. ellipticus
Antennaria microphylla
Cerastium arvense
Not gooseberry
Draba
Zigadenus venenosus
Arabis drummondii
Delphinium nelsonii
Besseya wyomingensis
Teeny tiny yellow flower
Collinsia parviflora
Lupinus argenteus var.
argenteus
Marsh marigold
Globeflower
Willow
Buttercup
Glacier lily
Spruce
Windblown pine
Leeks
Kinnikinnick
Yarrow
Bromus tectorum
Hackelia floribunda
Chrysothamnus nauseosus
Ribes coloradense
Vaccinium membranaceum
Lewisia rediviva
Linaria dalmatica
Opuntia polyacantha
Sphaeralcea coccinea
Sage
Rabbitbrush

# Preface and Acknowledgments

This book is a record of transformations both of my subjects and of my own. Its chapters are experiments. They tell stories in different ways, and their order is another story, in academic prose giving way to narrative and memory, about the ways we learn from (and about) people in specific places. A few simple facts led eventually to a long scholarly and personal project: I live and work where my subjects did, they were botanists, and I like plants. Each of these facts has a history, which in turn allowed other connections to happen between my subjects and me.

Aven Nelson (1859–1952) was a botanist at the University of Wyoming; his second wife, Ruth Elizabeth Ashton (1896–1987), much his junior, was his partner in the field and companion during the last twenty years of his life. These naturalists brought work and play, learning, pleasure, and companionship together fairly seamlessly in their lives. Neither of the Nelsons wrote about what it was like to work outdoors, what the plants they collected made them think about, or what it was like to work together, though it is clear enough they enjoyed themselves, and both of them sustained long productive lives in close communication with the natural world. Their lives ended long before I came upon their documentary remains, and the Nelsons remain mostly strangers to me, experts in a field that is not mine. But they lived in a place I know, and worked as botanists in meaningful and enjoyable ways which ultimately changed my own work in turn. Botany in the Rocky Mountain West is the fulcrum of my relationship to these people. Studying and exploring their experience of botany allowed me to learn *from* them rather than merely *about* them.

I became aware of the Nelsons on a visit to the Rocky Mountain Herbarium at the University of Wyoming shortly after I arrived on campus in 1997 as a new faculty member in American studies. I had written to the botanists at the herbarium in the early 1990s, when I was studying with weed ecologist Bruce Maxwell at the University of Minnesota. At that time I was writing about the introduction and spread of weeds in the American West for a chapter of my dissertation. The dissertation was a cultural (and critical) reading of western agricultural history, published in 1996 as *The Culture of Wilderness: Agriculture as Colonization in the American West*. Visiting the herbarium in person was an opportunity to connect graduate research (which I had enjoyed very much) with my new place of employment (about which I knew almost nothing). The herbarium curator said I might be interested in the biography of the herbarium's founder, written by a former University of Wyoming history professor. Reading Roger Williams's *Aven Nelson of Wyoming* gave me the idea that I could make my study of agricultural history more concrete by writing about the work of an individual scientist who had, in Nelson's case, taught at a land grant college, done research for the U.S. Department of Agriculture, and founded an institution (the Rocky Mountain Herbarium) that represented an archival effort related to those undertaken by many federal agencies (of which I had been very critical in my first book). I think I expected information about Nelson's life and career to fill in—at a fine scale—the fairly sweeping themes I had worked out in the book; not especially thoughtfully, I considered information about Nelson as data, merely the stuff of academic study.

But it didn't take long before Nelson became compelling as a person I wished I had known. He had made a whole life in Laramie, at the university, and in this particular landscape, where I was still struggling as a newcomer. Hardly equipped to examine Nelson's accomplishments with critical distance, I realized (with delighted surprise) that I liked Nelson. He was a modest, soft-spoken man, a beloved teacher, a loyal husband and father. As a fellow faculty member at the University of Wyoming, I admired his professional success, which it seemed he had achieved against all odds, both institutionally and geographically. I became curious about him the way one is curious about an acquaintance, not a research subject. I wanted to know where his house was, and how his wives had contributed to his work (later realizing I also wanted to know how he had contributed to Ruth's work). I was curious about what the university had looked and felt like to him when he first arrived, one-hundred-ten years before I did. To me, it felt like he had not really departed, though his grave in the local cemetery was real enough. Without fully realizing it at first, my encounter with Nelson (and later his second wife, Ruth) was shaped by the desire to have known him, not to study his work. That outlook became more and more conscious. I was finally glad to give up a certain kind of academic inquiry to acknowledge the fact that I had nurtured a relationship with Aven and Ruth Nelson, and through them, a relationship with the landscape in which I lived, and the university for which I worked.

Relationships are central to this book. One thing evokes and illuminates another, one person connects with another. New ways of living and learning can emerge in relationships, even with people one can't know well. These relationships in turn *take place*; objects of the natural world and the physical location of a life are each a nexus of human relationships, a medium for communication, memory, and shared experience.

This book is not an argument for the Nelsons' historical significance, or an exposition of the everyday work and thought of botany. It is not a biography or a history. Even as a series of prose chapters, the book is more like a collection of poems than an intellectual exposition of a subject. Form, voice, and image matter here; they are the indirect subjects of the book, even (probably especially) when they fail. I took risks in writing—some calculated and some reckless—because the subject seemed to demand it; I tried in form what I could not express in expository prose. The book has a narrator, not an author, whose voice and perspective have changed by the end of the book. Each chapter is about transformations of various kinds. Together they document an overarching transformative encounter whose occasions were vivid and precise, but whose experience can sometimes only be suggested, reiterated in microcosm, a varying theme.

"Work in Place," chapter 1, describes in an academic voice the contexts in which Aven Nelson developed as a Wyoming botanist who intensely desired to communicate his own appreciation for the living world to a broad public. The second chapter, "Specimens," looks into the signature artifacts of botany—pressed plants—for the capacity these objects have to pool memory. Though the voice remains recognizably academic, this interpretation of botanical objects, from a point of view that is decidedly neither botany nor historiography of science, begins to reveal a moment when strangers meet.

The material that follows—an album of fragments sorted by astrological signs, a series of letters, and the chapter, "Habeas Corpus"—faces the problem of writing about the liveliest of artifacts, a human being, Ruth Ashton Nelson, though she left few records. Her husband's extensive records can give us the illusion that we know a lot about him; hers invite more self-conscious engagement with the imprecision of knowing any person. I have made this fragmented material more strange than it is in its original form, especially in the album, partly to make a point, partly to honor the fact that this woman was alive in distinctive ways even if we can't know her. Whatever a reader thinks of the album, to me this was a gesture of gentle and thoughtful regard. My familiarity with Ruth Nelson is contrived and partial, but still approachable. The information available about her, or in her own hand, did not lend itself to consolidation in a grand exposition of her character or accomplishments. The fact remains that I recognized and responded to aspects of her experience that resonated with my own. Relationship emerges here as a significant key to memory, identity, and learning.

The last chapter, "Collecting," is a layered narrative in two voices that speak to each other as well as to all the subjects of the book. "Red Desert Reprise" is a coda.

What began as a quiet reluctance to study the Nelsons gave way to an explicit desire to write about my search for them after I spent the summer of 2001 approximating some of the Nelsons' field experience. I retraced their routes in week-long or ten-day trips around Laramie and the Snowy Range in May and June, in Yellowstone and Grand Teton national parks in mid-June, in Wyoming's Red Desert later that month, back to the parks in early July, and to Denali National Park in Alaska at the end of the month. I wanted to find plants that the Nelsons found, and write down what it felt like to be in these places looking for plants. I had to learn some botany; I practiced at home, took botanical books with me on the road, and made a press to try my hand at pressing plants (outside the parks where it was legal). I paid attention to what it was like to concentrate on plants under conditions of fatigue, elation, hunger, delight, thirst, pain, fear, and surprise—and what the whole activity made me feel and think, especially in culturally or historically significant places, exactly what Aven and Ruth left no record of. They were up to more than just botany, even when they were "working," even if I could not know the full extent of their experience in the field.

Recording something of my own that could not possibly replicate their experience led me to the uncomfortable conclusion that I had to find a way to make a description of the Nelsons' botany more than a mere account of these two botanists and beyond the context of botany (or even field science) as such. It was impossible, first of all, to experience the places I visited without thinking about the Nelsons' companionship, compared to the fact that, until I went to Alaska with my daughter, I retraced their work alone. I had many hours to remember my grandmother, who would have loved the landscapes I walked through and the plants I saw, touched, and pressed. The choices and accidents that brought me to these places and this work by myself were the constant backdrop to my travel; so were the limitations of an education that had prepared me for something else (though I was beginning to doubt I knew what this was). Week after week I learned botany, let my life be taken over by movements and habits the Nelsons had cultivated all their lives, and persisted in looking at and identifying every plant I could focus on. But by the time I left the Red Desert I knew there was a larger experience at stake.

What I was (re)discovering was not the Nelsons in any direct way, but a physical, intellectual, and emotional experience of "nature" through relationships with other people—remembered, enjoyed in the present, imagined, partial, or indirect in books and papers—that the Nelsons probably understood, whose ends lie well beyond botany or any intellectual project, and also beyond even the most ecstatic "nature appreciation." Settling the matter of whether they recognized such an experience or not would have been speculative at best. That complex experience itself became the subject of a different kind of book. Describing and analyzing how human relationships of various kinds shape experience and knowledge of natural objects and places gave way to feeling and *inhabiting* this process. Historical narrative and scholarly analysis by themselves are not the most appropriate approaches to a project of this kind, and I think not rigorous enough for it. The result is that I know a lot more Rocky Mountain botany than I did, and that I found ways of working and living that are better—aesthetically, emotionally, intellectually—than the ones which originally brought me to this work.

People learn what they learn under specific conditions: with particular people, in particular places. I believe they learn about places through relationships with people, too. The process of learning and transformation I explore here isn't unique to learning about the natural or nonhuman world. People, objects, and places become lively in anyone's life in specific ways. Anything we learn—geology, popular culture, hunting, French history, carpentry, mothering—can provide a route to personal and historical understanding. Any work in any place with any set of objects can become the necessary framework for a person's identity and memory, allowing a person to know she or he is alive in a responsive world, and providing a wealth of images to think and live in. (Joan Richards's recent book, *Angles of Reflection*, makes a good case for mathematics in her life; her bravery was encouraging.) But this book does speak specifically to environmental experience in at least one way. When we worry about people's "attitudes" toward the nonhuman world, and their habits of living with it, it is too easy to implore people to recognize that "everything is connected to everything else"—to tell people how complex the world is, all the things they ought to know about it, or how beautiful it is. It seemed valuable to me to think for a while about exactly what a "connection" is, what qualities make a responsive and nurturing relationship of any kind, and then to *do* these things, not apart from scholarship, but in written work that is in the end more invitation and interaction than description and analysis. I assume living breathing writers and readers can do more than think, even through print. They can—and probably should—feel, remember, dream, and change, too.

I wrote in the preface to *The Culture of Wilderness: Agriculture as Colonization in the American West* (1996), that it "came from somewhere"—an informal PhD before the real thing, living and working as a ranch wife in Montana. This book comes from very close by; it's the effort to complete work with more honesty than I was capable of then. The family I didn't name in that preface is the Zimmerman family in Wilsall: Chloris and Bud and their sons, Ned and Van. Van is my ex-husband. Bud, who died in 2001, had raised Herefords on Daisy Dean Creek for fifty years; those cattle, and land sales from his family's ranch lapped up by growth around Billings, sent both of his sons to Cornell University in the 1980s. Chloris manages her father's neighboring Guth ranch on behalf of the extended family, and serves on the Park County Planning Board. Ned and his wife, Cindy, run the Zimmerman Herefords and commercial cattle now, and Van is a computer graphics designer and editor in Boston. Our daughter, Grete, who has family splashed across the country, spends as much of every summer as she can in Wilsall. I understand she knows how to open gates.

I drive her to Granny Z's in about ten hours from Laramie. It's a familiar drive now under a proverbially big sky: through the rolling hills of Shirley Basin mining and ranching to Casper, across the fraught rangeland of Johnson County, west over the Crow Reservation and past the Little Bighorn Battlefield to Billings, through the bright lights and smoke of Laurel's refineries (like those that defined Lackawanna for me as a kid in western New York), to where the Absaroka Mountains hold up the southern horizon, while the Crazy Mountains drape down from the north. Turn right before Livingston and the scale of the trip becomes dense, on the longest twenty-five-mile road Grete and I know. The first time I revisited Daisy Dean Creek and the Zimmerman houses after I'd left them in 1989, I remembered what an extraordinary place that was, how much I had wanted to live there. Van and I worked hard there. My story has settled over time into the shifting layers of stories of people I am grateful I still know. They are family.

Grete and I moved to Laramie when I accepted a job at the University of Wyoming in 1997. Given a choice after several nonsensical years on the job "market," I knew I wanted to be in Laramie. I have spent the last six years remembering and discovering why. It's not an accident that I chose to work where I do, or responded to the work of two obscure botanists who lived here before me.

The too-easy story is that, like millions of other people, I fled west, not once but twice. But it's more complicated than that. My mother's family fled west, too, from one dirt farm to another till they pooled up in Napoli and other small towns in New York's "southern tier," south of Buffalo, sometime before 1900. My grandmother left the farm for teaching, and married in 1922. My mother, Joanne Puccio, secured the family's first college degree (in teaching) and conventional middle-class marriage (after a false start). I knew that farm as a little girl. My great-grandmother, Ida Pearl Ellis, worked it till she couldn't any more, about two hundred acres of woods, cornfields, pasture, and gardens. She drove a monstrous black Packard from the 1940s and grudgingly allowed her house to be wired for electricity and plumbed—in the kitchen—around 1960. She had a toilet then, too, behind a curtain in a back corner of the kitchen. She preferred the outhouse. She cooked on a wood stove (the whole house was heated with wood), and grew ginseng, among other odd things I did not get the chance to ask about. I took her leafy flowerless beds and the coiled metal handles for the stove burners as natural facts of all old people. I still have the wooden bowls and iron slaw chopper her husband made in what seems like another world. She'd been blessed by Billy Sunday, loved sentimental nicknacks, and was remembered as a battle-axe.

That part of New York, not just the living landscape (maples, sumac, trailing arbutus, and so on) but the time warp, was where I spent as much of my childhood as I could, with my grandparents, in a deep lot heaped into roses, tomatoes, lilacs, corn, peonies, the great bounty of living in one modest place for a very long time. My grandmother, Hazel Fargo, taught the entire town of Randolph fourth grade by the time she retired after fifty years; janitoring in the school was eventually my grandfather Harold's permanent job, which lasted long enough for him to die of emphysema from shoveling coal into the furnace. Every night of their lives, they wound their watches and laid them with their glasses on the dining room table before turning the fire down and going to bed. More than anything I have ever wanted, I have wanted to live there—not "there," New York, but there in the folds of real people and things, memory and stories, the voluptuous Everything outdoors, including what had ever happened there in the houses and landscapes occupied decade after decade by beloved people. My grandmother died when I was a junior at Cornell. In a flurry of grief, I took furniture and smaller things I wanted back to Ithaca. Dressers and chairs. Her aprons. China and silver. Photographs. Her school notebooks. All her kitchen utensils (some of them her mother's). Her house was sold, and I still dream in it twenty years later.

Not strangely, I rearranged my life by the time I graduated. I would not go to graduate school after all to study Milton; I couldn't write a "statement of purpose" for an application. Van and I were energetically in love; he was physically graceful, a fine writer, whose sensitive being in the world still moves me. I went home to Montana to meet his family in December of our senior year. Looking long into the old house that stood empty on the Strickland Place, part of the Zimmerman ranch, with our arms around each other, I knew I wanted to build a life with this man. We moved after our wedding in May, and the Strickland Place became "there": the mouse-ridden house with its many repairs, old and new; the willow creekbottom full of the smell of deer; the barn where young mother cows sometimes backed their calves into corners; the battered fence pliers and irrigating shovel Van wielded with swift beautiful precision; the shithead range horses Van rode seamlessly and I struggled to make peace with; the new fence whose poles we cut and peeled ourselves; the vast 1969 Oldsmobile I learned to drive in; the loads of hay we stacked (or restacked when I tipped them off the truck); the roses I planted and irises I found while uncovering an older garden. Grete's middle name is a flower we loved to watch in spring, Anemone. She thanks God every day, I'm sure, that we didn't name her Daisy Dean.

The story of why we couldn't stay there isn't just mine to tell, but I can say that the experience of leaving to go back to school was the kind that reveals and rearranges whole characters permanently, including my own. I should have paid attention. When I left, maybe because I was still fairly young and had my hands full with a two-year-old and coursework, I thought I'd really left. Graduate school is its own time warp. I couldn't stay away from Wilsall in the end—I wrote a book with too many syllables about agriculture in the West, when what I'd wanted to do was growl and wail as much as make an argument. (I still can't read, much less write anything directly from the journals I kept in Montana.) Graduate school and the environment I lived in taught me nothing really about how to bring these things together, much less what other things I might hear or say. Studying and writing as if this were just work, barely legitimate work at that (I never was a historian), about a place and "place" I no longer lived in, was an exercise in brain- and character-damage. I wouldn't know how much damage until I unraveled my entire household, including a second marriage, in the weeks before I left Minneapolis with Grete for Wyoming.

The academic job was a fact I could sink my teeth into, but Wyoming was suddenly abstract. The sliver of the dreamtime that brought me here was no match for feeling lost. "Assistant professor" is not a life, or a place, or a memory; working toward tenure gives it illusory shape and keeps the bills paid. My daughter was nine and furiously unhappy—she'd lost all her friends and her mother was unrecognizable. That, too, I sank my teeth into soon enough. The rest was a wash. I tried to work on subjects—Wyoming, an herbarium, and a botanist—that spoke to a lifelong attention to living things, that might also tell me where I was, as if this "where" were a geographical place. One big dog, and then two, meanwhile turned the backyard into a corral.

What took me six years to realize, as we slowly transformed the scrappy little house that was ours to buy, as Grete and I made friends and became vibrant, and tenure was breathing down my neck, was that an education and a profession—for me, a detour—are dwarfed by much longer, thicker, and more satisfying stories that can absorb work in unexpected ways. Other people may finish their reckoning with memory, work, and place long before their thirties, but I am not sorry I took the path I did. Five hundred miles from the Strickland Place, sixteen hundred miles from my mother, my grandmother, and my great-grandmother, what I began to learn "there" I have here. Wyoming poured into my life through my work. It comes into my head, my walking, digging, and planting, my writing, my memory, and into Grete's summer migration north a ways and back. This is home. There is, I hope, a long future to build work of all kinds here.

I OWE PUBLICATION of this book to Oregon State University Press editor Mary Braun's enthusiastic support and quick thinking—she sent the manuscript to Holly Carver at the University of Iowa Press—and I thank Holly Carver herself and editor Wayne Franklin for offering it an appropriate home. I was fortunate to have encouraging and helpful readers' reports from interdisciplinary literary scholars Scott Slovic and Susan Kollin. Slovic's work, editing the reflections of naturalists on their careers, through the Credo series, has been an important touchstone in rethinking my own relationship to professional work. Susan Kollin continued an invaluable exchange we began in graduate school at the University of Minnesota, in the Center for Advanced Feminist Studies.

I had the chance to try out an early version of this book at an unusual conference in 1998. Werner Sollors of Harvard University's American Civilization program (one of the oldest American studies programs in the United States) organized a gathering of "Young Americanists," presenting the work of nine recent PhD's in American studies. Each of us gave a formal lecture followed by a prepared response, in my case co-authored by Jessica Dorman and Steven Holmes. Holmes's book of and about environmental life-writing, *The Young John Muir*, would be published shortly afterwards. I was wondering at the time if the best approach to Nelson might be through systems theory, looking at the way that "small inputs" (in the form of biographical minutiae) might have large effects in certain systems (in the form of regional, institutional, and scientific developments). Holmes and Dorman focused on the life-writing effort and eventually so did I.

A University of Wyoming College of Arts and Sciences Faculty Grant-in-aid in 2001 funded my summer's research looking for the botanists who were the occasions for this book, and the University of Wyoming American Studies Program generously paid for photocopying and the reproduction of images used here as illustrations from the American Heritage Center. The archivists at the University of Wyoming American Heritage Center moved mountains of boxes of the Nelson papers for me, and carried a photocopying bill a mercifully long time. The director and curator of Rocky Mountain Herbarium, Ron Hartman and Ernie Nelson, have been welcoming with their time, expertise, and space since I started looking into their work and Aven Nelson's botanical legacy in 1997. Nelson's biographer, Roger Williams, retired professor of history at the University of Wyoming, shared many hours of conversation with me about Aven Nelson, the history of botany, Wyoming, and the university. Photographer and friend Michael Tierney confirmed the lure of pressed plants when he came to Laramie to work with Rocky Mountain Herbarium specimens for a week in 2000. Some of his photographs of those specimens appear in this book. Spiritual journeyman David Lang suggested what quickly became one of the most important sources defining the outlook of this book: Frances Yates's *The Art of Memory*. Rocky Mountain National Park librarian Sybil Barnes generously located documents and contacts for me regarding Ruth Nelson's life and work in the park, and gave me directions to what had been Ruth's property above Estes Park, Colorado. Jim Farrell, a former colleague at St. Olaf College, the best mentor imaginable for someone just out of graduate school, and now a good friend, provided the title for this book through his energetic and inimitable wordplay.

My colleagues in American studies, women's studies, and environmental studies at the University of Wyoming have lent their support and encouragement through the tenure process as the manuscript of this book was in tense limbo. For reading every scrap of my written production over the last six years, and making the job I came to do here the best possible job for me, you all have my enduring thanks: administrative geniuses Eric Sandeen, Cathy Connolly, and Harold Bergman; colleagues and mentors Ron Beiswenger, Gregg Cawley, Barbara Chatton, John Dorst, Jeanne Holland, Phil Roberts, and David Romtvedt. Bright and engaged students heard about this project from beginning to end, read it, and talked to me about it, as they lost and found things in their own research, writing, and teaching: Rob Chester, Andrew Grace, Richelle Lucas, and Cinda Nofziger. Thank you also, Sophia Beck, our incomparable office associate, for humor, wisdom, paperwork, and prayers.

A strong matrix sustained me while I was writing. In Laramie: Kathleen Harper, Cherie Lowenberg, Ellen MacQueen, and Peg Nelson. Farther away, close friends from graduate school: Rachel Buff and Lisa Fischman. Rachel introduced me to fluent and ambitious astrology when we were neighbors in Minneapolis, in the years when she plotted a new journal, *Marxist As-*

*trologer*. Lisa heard and read all of it, even when I hadn't said or written it all yet, and helped to reckon its roughest edges (and mine) month after month. She also came to Laramie to help me finish the last revisions when the press deadline coincided with the arrival of a new baby girl.

My mother, Joanne Puccio, gave me a biography of Orra Phelps, which helped me make sense of Ruth Nelson and my own relationship with my mother, her mother, and her former mother-in-law. My mother's love and stamina are unwavering. My father, Frank Knobloch, introduced me to Gregory Bateson when I was barely old enough to understand something interesting was up in those "metalogues," but I returned to Bateson much later, including here on the subject of empathy and recognition. My brother, Scott Puccio, fed the dog and kept the house from burning down while I did my field work, and found his own niche in Laramie. My father's sister and brother-in-law, Betsy and Bill Robertson, hosted Grete and me in Fairbanks, and provided warm sleeping bags.

Grete Zimmerman is a beautiful living reminder of what learning, joy, and companionship have to do with one another, that the world ought to be a better place, that a roof over our heads and kibble for the dogs are fine things, and that no work should be debilitating. Craig Newman, who asked me what I was doing on his desert in 2001, lovingly let me out and let me in. With him, Grete, and little Nellie Grace, I know that each life is a promise and a road into beloved country.

Anemone patens L.
Det. C. L. Porter 1959
Pulsatilla ludoviciana (Nutt.) Heller.
A. N. Dec. 11, 1900.
FLORA OF WYOMING. 2
Anemone patens nuttalliana Gray.
Telephone Canon.
Coll. Aven Nelson. May 29 1894.
Herb. University of Wyoming.
ROCKY MOUNTAIN HERBARIUM
3909
UNIVERSITY OF WYOMING

Reed grass
Flax
Fringed sage
Wild roses
Little yellow mustard
Rayless erigeron
Phlox longifolia
Allium
Sedum lanceolatum
Linaria dalmatica
Phlox multiflora
Descurainia?
Anonymous
Achillea millefolium
Linum perenne var. lewisii
Delphinium nuttallii
Taraxacum officinale
Yellow composite
Castilleja
Oenothera caespitosa
Tragopogon dubius
Melilotus officinalis
Salix
Astralagus
Erigeron acris?
Eriogonum umbellatum
Castilleja sulphurea
Achillea millefolium
Phlox longifolia
Besseya plantaginea
Artemisia tridentata
Pinus contorta
Fragaria
Oxytropis campestris
Phacelia sericea
Ribes viscosissimum
Vaccinium scoparium
Buttercup
Ribes
Castilleja
Aquilegia coloradensis
Aquilegia caerulea
Tiny purple violet
Potentilla arguta
Astralagus spatulatus
Arctostaphylos uva-ursi
Sedum lanceolatum
Cerastium arvense
Juniper
Huge purple erigeron
Senecio canus?
Delphinium nuttallianum
Clematis hirsutissima

"Just as change stimulates us to look for more abstract constancies, so the individual effort to compose a life, framed by birth and death and carefully pieced together from disparate elements, becomes a statement on the unity of living."
—Mary Catherine Bateson, *Composing a Life*

# Work in Place

This is a brief study of the role of place in shaping intellectual work, particularly scientific field work about that place, in an individual life.[1] The work included botany and public advocacy for people's knowledge and pleasure in the nonhuman world, especially where they lived; the place was and is Wyoming, including the University of Wyoming in Laramie; and the individual was Aven Nelson (1859–1952). Nelson began collecting specimens for the University of Wyoming at the turn of the last century when laboratory work held greater prestige, creating a career and a regional herbarium in a location that remains on a far edge of American academic life. Nelson's geographical isolation in the West prevented him from securing a prestigious education early in the twentieth century, and the direct experience (much less love) of nature at home lay outside formal botanical education. Nevertheless, his fieldwork and teaching in Wyoming became the heart of Nelson's own understanding of his science, and deeply shaped his beliefs about sharing scientific knowledge with the public as well as his students. Intellectual work can be shaped permanently by place—institutionally, intellectually, and emotionally. This attempt to show, in one botanist's life, how that might happen raises a few questions historians might ask about western institutions and the careers that form (in) them. It also serves as an invitation to ask such questions about ourselves.

Locating Nelson's (or anyone's) work in a place demands a few words about both place and work. Place matters—so we've been told by environmentally sensitive writers pleading for a wider appreciation of the particularities of "place," natural and social (often both). Their hope is to stem the tide of thoughtless transformation of singular places into no-place, generic expressions of consumer "culture" and urban sprawl, without memory or intimacy, community or diversity, human and otherwise. Terry Tempest Williams, Linda Hasselstrom, Gary Paul Nabhan, and Wendell Berry, among others, excavate their home places for us, their emotional responses to nature and community, their childhood and adult reckonings with "place," guiding us to the natural and social peculiarities we're likely to overlook in our own backyards. Some of them moving, many of them sentimental or nostalgic, a whole life that includes paid work seldom enters these accounts. (The notable exception is Wendell Berry, whose corpus is an extended paean to work in place.) For most writers on "place," the paid activity that ties us to the commercial forces overwhelming the nonhuman world is what distracts us from place, or damages places, so much so that environmental historian Richard White had to argue strenuously in *The Organic Machine* (1995) that people's work could be a source of knowledge about and intimacy with the nonhuman world in specific environments. White castigated an entire environmental tradition with the very title of his essay, "Are You an Environmentalist or Do You Work for a Living?" The work in question is manual work: fishing, logging, mining, and agriculture.

Intellectual work "takes place" too, though we don't know much about the relationship between this kind of work and a full reckoning with place—the limits, the opportunities for curiosity and emotional response it engenders. Academics are rewarded for the prestige of their degrees and their fluency in the languages of abstract "fields," not their full habitation (including work) in places. When home institutions lie low in the hierarchy of academies, scholars have an incentive to inhabit the fields of their colleagues with more sense of accomplishment and connection than they may feel at home. Whole careers are made routinely from material that has nothing to do with home—in fact, the farther away from home and everyday experience the better (someone else's home and everyday experience are of course fair game).

We do know something about "field work," the intellectual work certain scientists do in specific places, including natural ones. Scientists' field work is a valuable source of information on the physical and biological character of any place. But writers on the subject of field work have been preoccupied with the meanings of field science and experience, not the place(s) where these things happen. "The field" is not assumed to be a permanent home—it is an exotic or removed scene of a special kind of science, having something in common with travel, tourism, colonial administration, and the purposeful and accidental confusions of identity and status that accompany such ventures.[2] "Place" may shape intellectual work in the field, but accounts of how it does so serve an understanding of large patterns of science, and ignore the constraining and inviting powers of those specific environments, including *institutions within those environments*, to shape whole lives as well as scientific insight. The outlook of Harold Dorn's *The Geography of Science* (1991), which explores the relationship between environments and the sciences that developed in them, is somewhat more useful in this context: "Science . . . changes over space and in the context of environmental conditions, and in some situations its development has as much to do with geography as history."[3] But Dorn documents mainstream science on a large scale (mathematics, engineering, astronomy, for example, in classical high cultures), and much is lost on the scale of understanding the local, the individual, and the marginal or ephemeral, the scale of this particular study.

Aven Nelson was born in Iowa in 1859, and became a member of the first faculty at the University of Wyoming when it opened in 1887. He founded the Rocky Mountain Herbarium there in 1893, and enjoyed a long career as a botanist of the Rocky Mountain region. He was remembered as a "lifetime prof.[essor] of botany," who "botanized widely over the Rocky Mountain region building the important herbarium at Laramie,"[4] one "whose lifespan as an active field botanist exceeds that of all other outstanding western botanists."[5] Locally, his now-elderly students remember him as a gifted teacher, and Senator Alan Simpson and his brother Pete Simpson discuss Nelson as a significant figure in Wyoming history in their spring course at the University of Wyoming, "Wyoming Political Identity." Nelson curated the Rocky Mountain Herbarium for fifty years, served as university president from 1917 to 1922, and as president of the Botanical Society of America in 1935, the first national officer of that organization to be elected from the Rocky Mountain West. His first wife, Celia Alice Calhoun, and their two daughters accompanied his botanical exploration of Yellowstone Park in

1899, where Alice maintained their camps and helped to press and dry plants. After Alice's death in 1929, Nelson courted his second wife, Ruth Elizabeth Ashton. He was seventy-two. A master's student in botany then working on a field guide to the flora of Rocky Mountain National Park nearby in Colorado, Ruth was thirty-five. The couple married in 1931 and enjoyed nearly two decades of companionship and collaboration before his death in 1952; after a long career of her own authoring field guides and popular articles, she died in 1987.[6]

Speaking as a senior member of his profession to the Botanical Society of America in 1935, Nelson said that he had been educated for "teaching and administration" in public schools before he came to Wyoming to teach English. Nelson quickly found himself "slated for Biology, a field in which [he] had no training, except," he said, "a boy's unsatisfied curiosity in regard to the native flowers that grew in the ravines and on the clay hillsides of the open forest of oak and hickory." Zoologically, his "training had been of the most practical sort," chasing squirrels, rabbits, and game birds with his dog and a long-barreled Norwegian shotgun.[7] Nelson's remarks to his colleagues intended to show how little he knew in the 1880s, but the boy's curiosity and his experience with well-known flora and fauna in Iowa, where he was born, were good training for his later work as a field botanist who was a remarkable teacher and respected naturalist in Wyoming. Looking back over a successful career in 1935, he could afford to admit a humble beginning. He had carved his prestige from a profession that grudgingly recognized the small western institution he helped to build and the underprivileged field work that gave him his professional identity and knowledge of home.

Institutionally, Nelson faced serious obstacles to professional development for decades after he arrived in Wyoming. His difficulty began immediately, and hinged on the unforgiving geography of expertise that privileged certain degrees from specific (primarily eastern) institutions for emerging specialists of any kind by 1900. Nelson was hired to teach English, but university president John Hoyt had inadvertently hired two people for the same post, so the job went to a man with a Dartmouth MA in English (who left no mark in his field). Nelson had only a degree from Missouri State Normal College.[8] He taught biology, but also geography and calisthenics, in a "university" of six faculty members that still had to produce its own high school graduates before it could begin the business of postsecondary education. The campus was a single sandstone building amid acres of native prairie.

"Prexy's pasture," the manicured quadrangle now in the middle of campus, was then good habitat and hunting ground for sage grouse.

Hardly optimistic about the future of this institution, Nelson eventually asked for leave without pay to get a graduate degree in biology at Harvard in 1891–1892, hoping not to have to return to Laramie as a permanent residence. While he was away, the university's horticulturalist Burt Buffum collected plants of all kinds, preparing the forage plants for Wyoming's (award-winning) display at the Chicago World's Fair in 1893. Nelson faced the job of identifying Buffum's leftover plants on his return from Harvard with an MA.[9] Nelson knew nothing about plant taxonomy—he'd studied biology, not much botany, though it's questionable whether he would have acquired formal field and taxonomic instruction even if he had studied more (a point to which we'll return). He identified these plants by using a small collection of botanical books, and stored the specimens, planning to add others from Wyoming and eventually Colorado and the central Rocky Mountain area (not to mention more books to the university's library).[10]

What had begun with an odd assignment took a simple turn: he really enjoyed this work. He began his own collecting in 1894, and looking back years later he said, "more and more material became my prime desideratum," and "the things I did when I could do as I pleased were field and herbarium work."[11] The university trustees officially recognized his growing collection as the Rocky Mountain Herbarium in 1899. His unexpected joy in the field and herbarium in Wyoming was the accidental beginning of Nelson's career as a botanist, and at the same time the kiss of death for his ambition to work elsewhere.

As he became more involved in the tasks of field collecting and herbarium organization in Wyoming (alongside his other duties, not to mention family life with two young daughters), he unsuccessfully sought new positions. One of his Harvard instructors, William F. Ganong, informed him frankly in 1895 that he could go nowhere without a PhD.[12] Residency requirements at prominent botanical schools and the necessity of keeping up with all his work at home made the PhD an uncertain goal at best. The work that gave him the most pleasure in Wyoming was also difficult *because* he was in Wyoming. Plant collection was easy enough, but identification and publication were not. Nelson had no authority whatsoever as a botanist through most of the 1890s.

The bright lights of botanical knowledge were far from Wyoming—

Benjamin Robinson at Harvard, Edward Greene at Berkeley (and after 1895, Catholic University), and his rival and contemporary in western botany, Per Axel Rydberg at Columbia and the New York Botanical Garden. These men enjoyed not only associations with prestigious institutions, but access to large, well-established herbaria on the East Coast, the backbone of taxonomic plant study. Nelson's MA allowed him to stand credibly before classrooms of botany and zoology students, and elementary teachers in training, but that was all. As he struggled to gain professional standing after he began collecting in 1894, he had nothing like Harvard's Gray Herbarium at hand where he could compare plants he found with other specimens that had already been named by genus and species, as Benjamin Robinson never failed to remind him.[13] For years Nelson relied on the competing determinations of botanists at more prominent institutions as he went about his work in Wyoming. Worse, his rival Rydberg was a "splitter"—he tended to assign new genus and species names to plants that Nelson (a "lumper") would have recognized as varieties of known species.[14] Rydberg's position at the New York Botanical Garden by itself could easily trump Nelson's expertise regardless of his burgeoning knowledge of and immersion in Rocky Mountain flora.

Still, in 1898 Nelson began to publish articles on the plants he collected.[15] Though botanists regularly contest each other's plant identifications, some of Nelson's earliest "discoveries" still stand, including *Phlox multiflora*—a common ground-hugging plant in alpine prairie whose white flowers are among the first to open in spring. John Coulter rewarded Nelson's familiarity with Rocky Mountain plants by choosing Nelson to revise his 1885 *Manual of the Plants of the Rocky Mountain Region*, a long task Nelson began in 1901.[16] This was the only professional break Nelson ever received.

The University of Wyoming remained poorly funded and administered while Nelson dug himself hip deep in botanical research. By 1900 he was very busy acquiring knowledge and plants but also advising Wyoming residents about municipal beautification, gardening, tree planting, and agriculture in his teaching, traveling, corresponding. Through an arrangement with a former Laramie friend and fellow member of the Methodist church who became a faculty member at the University of Denver, Nelson finally received his PhD in 1904 for a portfolio of articles he had already published.[17] PhD in hand, he published his revision of Coulter's *Manual* in 1909, but by then in midcareer, he gave up seeking other positions. He sustained his reputation through building the Rocky Mountain Herbarium, publishing species new to science from the Rocky Mountain West, and advising the Wyoming public on planting for beauty and agriculture. Given the haphazard (even illegitimate) nature of Nelson's academic credentials, and the institutionally remote location of his work, Nelson wrested a slim prestige in his field from the outer darkness of Wyoming.

Wyoming was *not* an outer darkness of field experience, however. The one advantage Nelson had, the very thing his eastern mentors and rivals lacked, was his access to—an overabundance of—the field. His home institution and his whole life lay smack in the middle of it. And he loved being in it. The most prestigious botanists were not active field collectors, even if they had established their botanical careers that way; some, like Rydberg, specialized in western flora expensively far from home. Asa Gray's responsibilities as a laboratory taxonomist and administrator eventually left him too busy to collect at all.[18] The Rocky Mountain West provided the only riches Nelson had in creating a career and a regional herbarium. In addition, Nelson's duties as a teacher (particularly of younger teachers) called on him to describe the purpose of scientific training and excite young people in botanical study. This would have been much less pleasant or effective had he approached it with abstract scientific discipline and distance from the field, much less disdain for his location. Working for a land grant college dedicated to public service and coeducation, most of Nelson's students would be children of his Wyoming neighbors, only a few of them destined for more elite institutions. He did seek "better" appointments for a while, but entrenched in Laramie, he embraced his marginality twice over: Nelson was a sincerely committed member of the UW faculty till his retirement in 1938, while he plunged whole-heartedly into primary field and taxonomic work that his profession as a whole left behind. His position at Wyoming in fact allowed him to discover, express, and teach values of field learning he might easily have been distracted from elsewhere.

To appreciate what was at stake in Nelson's deepening commitment to fieldwork, it's necessary to understand the direction scientific botany had taken by 1900, unfortunately for Nelson just as he began to like the idea of himself as a botanist. Textbooks of Nelson's day were ambivalent about the value of field study, privileging "closet" and laboratory work over basic field collecting and the intimacy with known places this work might facilitate. Botanical progress was measured by increasing attention to the physiology, morphology, distribution, and ecology of plants, much of which was studied in laboratories. Field experience for its own sake was an indulgence of school children, women, and other amateurs, and women in particular were understood as amateurs regardless of their training. Nelson's own second wife, Ruth, faced this limitation even with a graduate degree as late as the 1930s. Collections made in the field were to be studied at established herbaria (most of them in the East), not where they were found—not at the University of Wyoming—underscoring the prestige of a few institutions and laboratory study in securing the professional status of botany as a science. Everything outside the laboratory—including the outdoors itself—was beyond the scope of Nelson's own education and botanical education in general by the turn of the century. Botany focused on the abstract order and biology of plants, not the experience of finding and recognizing them in a known (much less enjoyed) landscape.[19]

Even if Nelson could be called a "taxonomist," taxonomy was a subspecialty by 1900. Taxonomy was where American botany began, in the collections of unscientific explorers curious about a new flora. Nineteenth-century American botanists had their hands full surveying and classifying western flora especially, and developed other branches of botanical science later than their European counterparts.[20] By the end of the nineteenth century, when Nelson began his own collecting, identification and classification were no longer the focus of botany and had become associated instead with amateurs. "Amateur and professional interests became increasingly incompatible during the 1880s and 1890s," argues Elizabeth Keeney, "as the dominant professional focus shifted from natural history to biology. . . . Unlike natural history, [scientific botany] employed experimentation as well as observation, stressing physiology and ecology over taxonomy." The emerging professional science—new botany, as distinct from untrained, amateur collection and classification—"did not abandon taxonomy, but de-emphasized it in favor of physiological and ecological issues."[21]

Taxonomy remained important, but subordinate. Taxonomy "is at once the alpha and omega" of botany, Reed Rollins wrote in a historical overview of American botanical taxonomy in 1958, though there was no confusing the two ends. About any organism, the "first question asked is: What is it? This was also the pristine question of man." But as a result of the development of scientific botany, when the question is answered, "one possesses an open sesame to the accumulated wisdom and knowledge of our civilization concerning a particular species. Important as it is to open the door, it is of far greater significance for taxonomy that it stands at the end of the line to profit from any and all inquiries that may be directed toward a given organism.

The numerous modes of research represented by the various branches of botany, ranging from physiology to palynology"—the proliferation of scientific, laboratory-based specialties Asa Gray had encouraged—"produce a mass of data about given plants that is not only significant for its own sake, but often is very pertinent for the systematist."[22] Taxonomic botany required a special eschatological defense by the 1950s in the context of newer sciences, even for Rollins, who was an acclaimed taxonomist. Rollins was the Asa Gray Professor of Systematic Botany at Harvard, the director of the Gray Herbarium, among the founders of the International Association of Plant Taxonomy, and one of Nelson's former students.[23]

By the time Nelson was even employed in Wyoming, the value of collection and classification had changed as a result of the rise of new botany. Some of it in the hands of amateurs, and within professional botany eclipsed by new specialties, collection and classification and the fieldwork on which they depended were at the margins of professional botany. Moreover, professional consolidation of botany as a science happened just as western institutions emerged as potential centers of new botanical knowledge. Such a bias would have undercut the ability of people with appointments and interests like Nelson's to be recognized for taxonomic work undertaken from home rather than in the service of what would have been, to them, remote eastern authorities. This was the inexorable tide of scientific botany against which Nelson struggled till about 1910, and significantly abandoned afterwards, at least as far as his career advancement was concerned. For Nelson, though, outdoor experience became the scene of botany's greatest purpose: encouraging students and the public at large to encounter nature where they lived with joy as well as knowledge.

It wasn't scientific botany but nature study that gave Nelson a legitimating intellectual context for the kinds of work that had come to occupy him. Those who promoted the study and experience of nature among elementary school children in particular understood that nature enjoyed outdoors was inherently interesting, and love for nature was a respectable goal for nature study in school, even if one moved onto other forms of knowledge in higher education. Part of Nelson's job was to educate future teachers, and he included nature study in their instruction by 1905. Nature study theory and lessons were developed at the end of the nineteenth century in response to the deadening laboratory emphasis of natural science education, in turn a direct result of the professionalization of the biological sciences. Cornell horticulturalist Liberty Hyde Bailey was its preeminent champion. Nature study promoted children's knowledge about nature and lifelong engagement with and respect for the natural world, with the hope that it would encourage an interest in farming and rural life. The new botany, which Nelson dutifully passed on to his college students in laboratory-based courses, was about microscopes and abstract problems, not the world in which botany's subjects were found. Nature study was adopted in high schools where the new botany, if it was taught at all, failed to ignite anything beyond students' boredom and frustration. Nature study was a new approach to scientific knowledge within public education, and by 1900 it was clear that professional botany had effectively divorced itself from the curious public.[24] The new curriculum, which stressed the lessons provided by students' local natural surroundings, was in part a revival of naturalists' work that had been abandoned in formal scientific training by the turn of the century. Liberty Hyde Bailey wrote in 1903, "Nature-Study is not science. It is not knowledge. It is not facts. It is spirit. It is concerned with the child's outlook on the world."[25] Though its aims were grand and diffuse, this outlook was local in practice, and saturated with affect.

Nature study assumed that the natural world, not scientists, had something to teach people. In Nelson's classroom, young teachers were encouraged to use students' household pets as an inspiration for learning, as well as a range of well-known plants, animals, and insects students might encounter every day. Even pests—tapeworms and the organism that caused trichinosis—were part of the living world students knew, and therefore occasions for teaching.[26] What students were familiar with firsthand, both indoors and out, was central to nature study instruction. "The striking thing about [nature study] lesson plans is the degree to which they were driven by the students' interest," Elizabeth Keeney observed.[27] Students' direct experience outdoors raised questions, which further study would answer and expand, about the identities and uses of local plants, or the ability of weeds to take over uncultivated ground. It "focus[ed] on the child's questions rather than delivering lectures. . . . Tips for teachers cautioned against too much instruction."[28] Teachers provided materials for indoor study of plants, including microscopes and manuals, but also terrariums and potted plants which students could nurture and observe on their own.

"Teachers still found that an occasional walk to 'botanize' was both pedagogically valid and a welcome break,"[29] giving students an opportunity to work on a school herbarium, as well as exercise. Including outdoor recreation in school was intuitive to Nelson, as it must have been to thousands of teachers, though they surely welcomed official recognition of its pedagogical value. Nature study also valued what poor, rural schools in particular had much of—immediate experience in a well-known landscape—rather than forcing teachers to attempt lessons for which they had neither the equipment nor in many cases the expertise. At seventeen, teaching public school in Missouri, Nelson and his class took Friday afternoons "off for 'nature work.'" He wrote, "I raced up and down the hills and ravines with my whole flock, in hot pursuit of the birds and flowers." He had "armed [himself] with a copy of Gray's *Manual* . . . the first and only book on systematic botany [he] had ever seen." The book was a "disappointment," because "Nothing had led up to the vocabulary [he] encountered," but no doubt that did not interfere with his Friday routine.[30]

Nature study proponents wrote vividly about field experience, including plant collection. One example is especially striking. William Whitman Bailey advised nature study theorists as a practicing botanist, and introduced his daughter to natural history using its methods.[31] He wrote a popular volume on botanical collection that is as lively and engaging for the initiate as Gray's *Manual* is daunting. No doubt Gray loved his work, but gifted as he may have been, he was evidently dull as a classroom teacher;[32] his botany was not intended to encourage anyone to go outdoors. Bailey's did. He began, "The study of Botany is rendered especially fascinating from the fact that so much of the work is performed out of doors. In every pursuit are required hours of recreation and exercise. A stroll in the woods is then, of all things, most enjoyable."[33]

Like other botanists, Bailey encouraged close observation: "one of the greatest educational uses of natural science" is that "it trains one to see and to think."[34] But his tone and emphasis were not didactic. He valued outdoor experience for its own sake, and pressed specimens for their beauty. Bailey lavished many pages on how to collect and press plants, including the equipment needed, how to dress, what kinds of pressing paper to use, and so on, topics conspicuously absent or severely abbreviated in other botanical textbooks, then and now. His recommendations were spiced with memory and anticipation of pleasure, excitement, and good company. Even the herbarium was not merely a resource for scientific study: "Apart from any direct utility, no one is wasting time who studies the wonders of nature for their intrinsic loveliness. The pursuit of beauty is educational in itself, and often a practical adaptation is found where least expected."[35] More than that, herbarium specimens were reminders of experiences in specific landscapes, and the companions who shared them:

> In looking them over, one sees not alone the specimens themselves, but the locality in which they were gathered. Many an incident in life, the memory of which has long since become dormant, will be re-awakened as by an enchanter's wand. He will tread the forest paths gay with flowers; he will pause in imagination for the nooning by some fern-laced spring; he will climb the mountain ravine where the blood-root and orchis bloom, or wander, full of speechless yearning by the ocean shore. Not only do the natural scenes return thus vividly, but the faces of friends who enjoyed the occasion with him. He is once more seated, may be, by a little lake on the mountain, in a garden of alpine flowers. Cool streams flow by him, and he picks the tart fruit of the cowberry. The world lies mapped at his feet, and the infinite heaven is above him. He hears the merry jest and ringing laughter and his heart becomes gay with the thought of those oldtime rambles.[36]

Purple as it was, Bailey's prose makes clear that acute pleasure, geographical familiarity, and companionship, as well as knowledge, were part of botanical study.

Nelson shared Bailey's sense of the natural world as a beautiful and edifying place. Like many of his generation and background, from "childhood up" Nelson classified himself a "nature lover."[37] The Laramie Basin and Snowy Range gave the object of his love some significant territory. But Nelson was not a writer of popular works on the subject; his publications described new species or reported the progress of the herbarium. His nineteenth-century rhetorical flourishes would have marred his ability to speak or write to twentieth-century audiences; he delivered sermons when he spoke, quoting scripture to the scientists, and evolution to the congregations.[38] He encouraged curiosity for the natural world to his students and the public at large. To his colleagues, he preached jeremiads against their neglect of the public. He was disappointed in his fellow scientists for their distance from ordinary people and the world of nature available to them.[39] It was his perspective as a teacher and his own exuberance about the living world close to home that informed his work and words on behalf of the public.

Nelson spoke about making the pleasures and knowledge of botany accessible to nonspecialists in three passionate incidental addresses, two titled

"Science and the Laity" and another on "Science and the State."[40] His references to teaching reveal the literally homely world that was the wellspring of his connection to his students. "Before my classes," he said, "I have often stressed the fact that to enjoy [plants] you must really know them personally in their homes and ours—where they live, what conditions surround them, what kind of relatives and neighbors they have." Learning to "know some plants in this intimate way you will necessarily want a name for each kind, but that is a secondary matter, even though it is indispensable. Taxonomy is not coordinate with the great fields of botany—Morphology, Physiology and Ecology—Taxonomy is the Service Department for all the other fields." Its primary purpose was to help "thousands to know and to love" plants. Francis Ramaley—professor of botany at the University of Colorado at Boulder, for whom the address was given—wrote studies that for Nelson had "the distinction of being accurate, interesting, and above all, of being understandable, which is more than I can say for some I could mention."

Nelson was not arguing against precision or progress in scientific knowledge. Still, knowledge was only part of an experience of the world that included joy in familiar places, too. He said, "I would not have less of science for science's sake; less of the endless but necessary detail of organized research; all these and more too, but," he added, "I would have besides the great outstanding facts of each science so presented (if it be possible), that the lives of men and women shall be fuller and richer because they have touched hands as it were with a few of the loveable creations and creatures of the great universe."[41]

In another address on the same subject, he lamented that a great tradition had been sundered: "We no longer have any naturalists or, if we do, they are sometimes justly and sometimes unjustly called nature fakirs. Darwin and Huxley and Thoreau and Agassiz and Gray and Fabre have left no successors. We do not even have botanists any more. We have bacteriologists, algologists, mycologists, pathologists, ecologists, pteridologists, dendrologists, agrostologists, histologists, physiologists, taxonomists and so on ad libitum." Where had all the great generalists gone? At least one, University of Nebraska botanist Charles Bessey, had "passed on to the flower flecked plains of the new Canaan and to the roseate, blossom-filled fields on the light-kissed hill-crests of the new Jerusalem." Evidently they could enjoy a beautiful living landscape in the hereafter. Nelson's contemporaries had abandoned their role—even a profound duty—to "touch the lives of the multitudes," leaving their science bereft of the vitality of the living world in which they lived and worked. As a teacher with that vitality at heart, he said, "the joy of seeing eyes, blind to Nature's truth and beauty, suddenly open and brim with comprehension and pleasure has been mine." Sentimental as he may have been, Nelson understood ordinary people's comprehension and pleasure in nature as the highest goal of his science.[42]

Nelson's written effort in the popular vein was to be a revision of his 1909 *Manual* to encourage "the redevelopment of interest in Botany among intelligent people everywhere" (traveling or living in the Rocky Mountain region presumably), a technically simplified *Dual Purpose Manual*. He wanted its use to inspire "that type of happiness that comes from a sympathetic contact with and an adequate understanding of that part of our environment that brings re-creation to our bodies, joy to our minds and peace to our souls —the world of life around us."[43] This would have been a tall order for any "manual" of botany, but Nelson's intentions were clear. The revision was never published.[44]

Nelson accomplished directly in formal and informal teaching what he couldn't in print and oratory. He taught generations of students and corresponded actively with a curious public. The public (particularly women) wanted information about Wyoming wildflowers and often birds as well. Corabelle Ewel, chair of conservation for Wyoming women's clubs in 1929, wrote to him asking for "anything that would be available for the use of the Club women for reference work" identifying birds and flowers. Tacetta Williams of Thermopolis asked him "whether there is in circulation a book on Wyoming birds and flowers, illustrated and described in common enough language so that any old sagebrusher can understand it." Rose Snell of Fort Laramie wanted to know if Nelson's *Manual* could be used by small children to recognize flowers. He had little material to give them, though Nelson invariably recommended his *Manual* (if not to school children), and hoped for the republication of a number of early pamphlets on exactly these subjects.[45]

Other correspondents received more detailed responses. Mrs. Charles Bigelow of Minnesota had visited the Big Horn Mountains and sent Nelson a pressed flower: "It is a virgin blue, and wonderfully lovely,—is it not? For you have recognized it at once I am sure." Nelson wrote back that it was "doubtless *Viorna hirsutissima*," or as he preferred, *Clematis hirsutissima* (sugar bowls, or vase flower—furry overturned cups with graceful points around the brim). She had tried to identify it using Rydberg's *Flora of the Rocky Mountains and Adjacent Places*. For her knowledgeable appeal, Mrs. Bigelow received Nelson's standard recommendation of his own book, with the explanation that he avoided Rydberg's "extreme subdivision of genera and species."[46] Mrs. Allen of Laramie asked Nelson about the possibility of protecting plants in Wyoming. He replied that "generally speaking our most conspicuous plants are holding their own very well," but he offered a list of ten candidates (with both botanical and common names), "sufficient to suggest that the wild flowers that add so much to our outdoor beauty are not able to withstand the inroads made upon them by thoughtless collectors."[47]

Nelson patiently answered almost all of these requests, and talked to the public about Wyoming wildflowers; his lectures packed the botany laboratory with standing-room-only crowds.[48] His correspondents and his audience were people who looked at things outdoors and wanted simply to know something about them, including their names. He understood he had a role in teaching the public as well as his students what variety of plant life thrived near their homes.

Nelson's greatest success conveying his knowledge and excitement for the landscape he lived in happened in the field itself, through the University of Wyoming science camp, established in Medicine Bow National Forest in the Snowy Range mountains above Centennial, about forty miles west of Laramie.[49] When the camp opened in 1923, Nelson was free from his duties as university president, and eagerly looked forward to new field work. To a colleague, Nelson wrote in 1929, "We are now prepared to furnish comfortable living conditions in a beautiful mountain setting in the midst of an unusually rich and attractive flora." He praised the quality of students the camp had attracted and continued, "If you should happen to know students who would be especially interested in what we are prepared to give and would at the same time enjoy the coolness and beauty of the mountains," Nelson would send information to them directly.[50] Botany students would be immersed in the field itself as an attractive and exciting place, hopefully kindling their interest and sharpening their skill in primary botanical work.

Begun by geologist S. H. Knight, its log buildings constructed by Knight and his students, the science camp drew college and graduate students from Wyoming and around the country to study geology, and soon botany and zoology, for several very productive decades.[51] Nelson taught there every summer until 1938. For once, the Ivy League came to Nelson. Many students enrolled from prestigious institutions like Harvard, Columbia, Radcliffe, and Smith. The camp housed a hundred students, about half of whom studied geology. There were usually about as many women as men, though predictably geology was dominated by men, and botany by women.[52] The science camp gave postsecondary students a field experience of immeasurable value.

After the first summer botany course in 1924, students said that they would long remember their experience "because of the very interesting work they were just finishing, because of the congenial bunch who were members of the class," but most especially because of their work "with the good fellow and true lover of flowers," Nelson himself. Meeting for the last time in his home, Nelson "told some of his experiences as a collector," as well as how he came to be a botanist in the first place. "Each person spoke a good word for the summer school. . . . The most enjoyable part of the work, it was agreed, was going forth into Nature's herbarium to study, not the dried specimens and a dryer book, but the living specimens in all their riot of color. To Dr. Nelson we owe the pleasure of becoming acquainted with these children of his, and for the lessons in seeing the beauty that is about us." The students were grateful for the privilege in studying with "one of the master Botanists of the time."[53]

Botany at the science camp was rigorous, but there were plenty of opportunities for fun, lyricism, and not a few mechanical hazards. Two botany classes in July 1926 "started up Telephone Canyon to spend the afternoon studying nature, and with the intention of eating a picnic lunch under the pines." A flat tire stalled the group. While some of the men worked at fixing it, Nelson led "the rest of the class through the pines and explain[ed] the differences between Pinus Murray and Juniperous Scopulorum, and that Pinus flexis is much greater than Eriogonum chrysocephalum even though one is likely to get tongue cramped trying to call the latter by name." For lunch Nelson opened a cooler full of goodies including "Honeymoon ice cream," which the students joked might have been the name of a new plant. The cars came home without incident, and in the end they said (with familiar student understatement), "it was a great party, and we learned a lot too."[54] Iris Harrington's response to that summer's work raised a joyful noise to Nelson's influence and her own deeply felt experience:

He showed me the souls around me in every nook and bower,
I saw the hand of Nature in every root and flower.
He turned my heart to a wavelength that e'er before was mute,

And I heard the Great Announcer's voice in sweeter note than flute,
As it spoke from every blossom, from all the grass, and trees,
Whispering love, and beauty, on every breeze.

. . .

And should I grow to a tree on the slope, or be dwarfed to a shrub on
the crest,
I shall know the same Life that exults in me, is the One that transforms
the rest.[55]

In both cases Nelson's students shared a warm connection with him, and obviously good memories of the field. These expressions of what they did and what they learned were "final exams" not for a classroom, but for a whole experience outdoors with this particular guide. Nelson's students enjoyed a rich experience of place, at home among Nelson's "children"—plants as well as his students themselves—which was no less formative for being fun. They turned botany to the service of humor and inspiration, including accurate references to local botany along the way. Many of Nelson's science camp (and other) students became accomplished botanists, including Wyoming-born Reed Rollins, whom I've already mentioned; he took Nelson's summer courses as an undergraduate in 1929 and thereafter majored in botany.[56] Some of the deep texture of what motivated Rollins and others lies in these exuberant memories of field experience with Nelson.

Nelson called the Centennial Valley "a botanist's paradise" before he was able to enjoy regular teaching there. "The lover of flowers need not go far in order to satisfy his longing for the beautiful in Nature."[57] Nelson was not writing about other people. Even if he did not write specifically about what field work in Wyoming gave him, people noticed he was in his element in the field with the companionship of his students. In 1927, children's literature writer Anne Carole Moore was pleased to meet Nelson, "the distinguished botanist who has discovered and named so many of the lovely flowers of our Western mountains":

> Here, at least, was a student of nature, who seemed to [me] as natural as a child out-of-doors. Armed with a kettle filled with smoking and smoldering twigs and boughs he was for the moment playfully protecting his student companions from the hordes of mosquitoes, which descended upon them as they sorted out the specimens gathered in the long day's tramp. . . . Very persistently . . . has the picture of the genial botanist of

Medicine Bow, with his kettle of smoke and his jolly looking students, stuck in the memory as a kind of symbol of what must happen before a truer appreciation for Nature . . . will become widespread.[58]

"At 70 years of age," another reporter wrote, "Doctor Nelson could still tire out his students in the field as he pursued his search up mountain slopes for botanical specimens, or climbed pine trees for specimen cones."[59] What became a lifetime of hiking up and down mountain slopes, sharing ice cream, shooing mosquitoes, and climbing trees was not just about systematic botany. The pleasures of fieldwork in Nelson's backyard animated his practice of botany.

Nelson's career took shape initially in the wake of the development of professional natural sciences distinguishing themselves from the amateur work of naturalists, shutting those with insufficient credentials out of professional mobility. Nelson's location in Wyoming, his sudden delight in botany, and his unabashed (if now dated) expressions of enthusiasm for nature, together gave him the motivation and the resources to use his science in the service of a broad public. His work opened the flora of Wyoming to himself, his colleagues, and the public. His was literally a unified field: the place of Nelson's institutional limits became the subject of his work, interweaving a new regional herbarium with extensive public service and abiding personal response in that very landscape.

Institutionally, Nelson faced limits that deepened his commitment to building the Rocky Mountain Herbarium. When Benedict Anderson described the looping arcs of colonial bureaucrats' relocations and promotions in *Imagined Communities* (1983) he argued that new nations—including institutions of nation-building, like censuses and museums—emerged in part from the artificial limits these functionaries faced in career advancement: they could not get work in the centers of imperial power, so they created nations and institutions where they were. The analogy to academic career advancement is not as far-fetched as one might think. Playing professional catch-up, Nelson's credentials kept him in the provincial scene of his work. His response, conscious or otherwise, was to create an institution analogous to those that would not have him—moreover, a regionally specific institution, which was something relatively new. If Harvard and the Gray Herbarium and its sister institutions in the East presided over all of botany as a general field, and Nelson could gain no admittance to the higher echelons of botanical prestige, he nevertheless produced a scaled-down model of his profession whose main virtue was that it was actually located in the region it documented. Moreover, with Anderson in mind, Nelson's work as a taxonomist was allied with the outlook of the "census," cataloguing the variety of Rocky Mountain plants; the herbarium was and is its "museum," a legitimating institution testifying to the unique natural history of this region, and the plausibility of studying in it.

Nelson was not the only man of his generation to use his position in the West this way. Consider the career of Frederick Jackson Turner. Born in 1861, educated like Nelson with the intent to teach, Turner hurriedly finished a PhD at Johns Hopkins University in order to keep his position at Wisconsin and release himself from teaching oratory, which he hated. Though he worked at Harvard, Turner preferred Wisconsin.[60] Is it a surprise that Turner would secure his intellectual place in history by defining a new field—western history itself—from a far edge of the historical profession, or that this place was reinforced, not undermined, by decades of controversy over his frontier thesis? We can see Anderson's looping arcs of advancement in both Nelson's and Turner's local public school educations, their efforts to secure master's degrees and PhDs under some pressure, even their detours through the "imperial center" of Harvard, and finally suitable employment—in and about the West. How many other western careers have taken similar paths? What western institutions or bases of knowledge do we owe to the accidental frustrations and desires of western scholars, "well-educated" and otherwise?

With respect to the intellectual content and practice of the natural sciences in particular, Nelson's career raises other questions. Did the development of field sciences at western land grant institutions in any way match the outline of Nelson's career? Many scientists at western institutions in the 1890s and early 1900s would have faced odd course assignments, extensive demands for public service, education of local young people, agricultural extension research for growing agricultural communities, and documentation of public lands, including national parks. Any of these things might have opened doors to rich careers grounded in locally specific environments in close communication with the public, though not necessarily acclaim as specialists. More generally, by 1900 there may have been two parallel institutions of scientific practice across the country (with or without regional variation): one in which high-prestige specialists were trained, and another more closely tied to the public, teacher training, and local communities and landscapes. If Harvard was the seat of American scientific botany (among other things), was the otherwise "ivied" Cornell the reference point of a different set of institutions more closely bound to nature study as a result of its land grant status and the work of Liberty Hyde Bailey? Historians of science have not focused on the professional geography of field sciences, though the split between professional and popular science is widely acknowledged and well documented.

Finally, we know and ask little as scholars about the role of affect, desire, and pleasure in work, our own or other people's. When the historical figures we examine "liked" what they did, or where they lived, that's nice, we might say, but it has little explanatory value. I disagree. In the context of environmental scholarship especially, understanding the emotional dimension of people's lives in places is I think crucial in understanding what they do there. Nelson didn't have to love what he did, or the place in which he lived. A slim spark of pleasure became a life, in what is probably an ordinary alchemy, tying his memory and childhood experience at home in Iowa to making a permanent home as well as a career in Laramie. The West, as a historical entity and a location of homes, recreation, and enterprise, is certainly saturated with many people's ideas about nature, the frontier, and the past and future. But it is not just "ideas" that motivate people, shaping both lives and places; for better and for worse, feelings do too, inchoate, unexamined, inconsequential because they are personal, irrational, and contradictory by nature. Places engage our affect, as well as our intellect, plying their way into work of all kinds.

University of Wyoming Campus, looking east, 1901. Courtesy American Heritage Center, University of Wyoming. S. H. Knight Collection.

Aven Nelson mounting specimens in his herbarium, 1914. Courtesy American Heritage Center, University of Wyoming. Aven and Ruth Nelson Papers.

University of Wyoming Science Camp lodge, built by geologist S. H. Knight and his students. Courtesy American Heritage Center, University of Wyoming.

Aven Nelson and students on a ridge in the Snowy Range, 1937. Note the vasculum he carries for plant collection. Courtesy American Heritage Center, University of Wyoming. Aven and Ruth Nelson Papers.

Geum triflorum
Anonymous
Potentilla fruticosa
Small yellow clover
Bright yellow trumpet
Geyser
Fumerole
Hot spring
Balsamorhiza sagittata
Rayless erigeron
Smilacina stellata
Taraxacum officinale
Fragaria
Succulent, not a purslane
Erythronium grandiflorum
Arctostaphylos uva-ursi
Fragaria
Pinus contorta
Lupinus argentus
Castilleja
Claytonia lanceolata
Potentilla gracilis
Dodecatheon conjugens
Trollius laxus
Nuphar polysepalum
Taraxacum officinale
Agoseris aurantiaca
Arctostaphylos uva-ursi

Viola adunca var.
bellidifolia?
Rumex
Polygonum?
Plantain-leaved buttercup
Spring beauty
Collomia linearis
Geranium viscosissimum
Delphinium nuttallianum
Phlox longifolia
Senecio
Lupinus argentus
Balsamorhiza sagittata
Collinsia parviflora
Viola nuttallii
Collomia linearis
Corallorrhiza maculata,
albino
Gilia aggregata
Ceanothus velutinus
Prunella vulgaris
Hydrophyllum capitatum
Lithophragma parviflorum
Eleagnus commutata
Amelanchier alnifolia
Mahonia repens
Potentilla gracilis

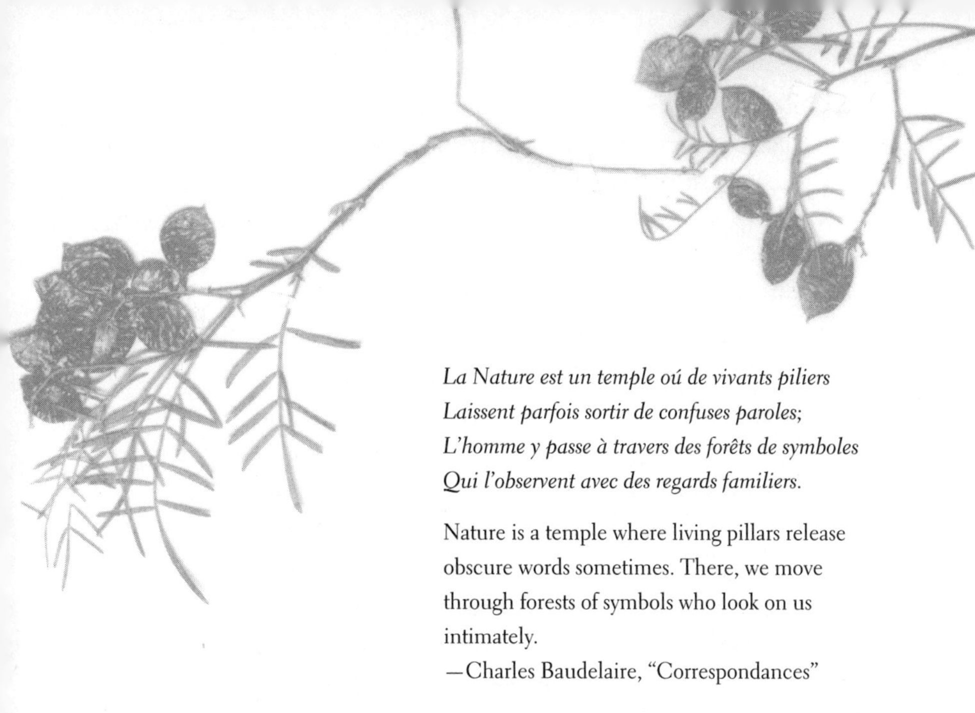

*La Nature est un temple oú de vivants piliers*
*Laissent parfois sortir de confuses paroles;*
*L'homme y passe à travers des forêts de symboles*
*Qui l'observent avec des regards familiers.*

Nature is a temple where living pillars release obscure words sometimes. There, we move through forests of symbols who look on us intimately.
—Charles Baudelaire, "Correspondances"

# Specimens

## CORRESPONDANCES

The Rocky Mountain Herbarium now occupies the third floor of the Aven Nelson Memorial Building on the west end of the University of Wyoming campus, near the oldest university building and Nelson's first workplace. It was built to house the university's library in 1922; its construction was one of Nelson's achievements during his university presidency. After a new library was built, in 1960 the botany department and herbarium collection moved out of the Engineering Building, where students (and teachers) still enter under the dictum, "Strive On—The Control of Nature is Won, Not Given." The staid block in which Nelson's botanical institutions and objects now reside is a stone's throw from the house where he lived at Fremont and ninth streets. The herbarium's director, Ron Hartman, walks to work as Nelson did. Over the door of this building is Aven Nelson's name. Its cornerstone is engraved: "READING MAKETH A FVLL MAN."

The third floor of the Aven Nelson Memorial Building is a hive of botanical activity and record keeping. Aside from a few offices and workrooms, most of the floor is one large room. Half of it is filled with herbarium cases, the original heavy wooden ones and newer metal ones alike perched on state-of-the-art movable tracks. The other half of the room is packed with worktables bearing maps, books, and cardboard boxes of unmounted plants; there is a long counter with microscopes under the windows. A photograph of Nelson hangs over a narrow shelf informally displaying Roger Williams's biography of Nelson, a sample specimen, a guest book, and some pamphlets, but a casual visitor would not really know what sort of place this was. The herbarium's riches, over 700,000 plant specimens from the Rocky Mountain West and around the world, lie stacked in the dark cabinets. A computer sits ready for additions to the herbarium's on-line database, an enormous project of data entry; when it's not occupied, the computer displays one of Nelson's more acerbic field observations as a screen saver: "a vile and most pernicious weed," it scrolls. The place is very much a library still. The herbarium has its own collection of noncirculating botanical books, shelved in one corner of the big room. All those cabinets likewise are a library of botanical history and knowledge, shelved by genus and species.

Chaos and order, past and present, reign together in this place. A box of sedges collected and somehow overlooked in the 1890s surfaced mysteriously shortly before I visited for the first time in 1997; they were mounted and filed into the cabinets by the end of the twentieth century. Field notebooks that would be catalogued and squirreled away if they resided in the bowels of the American Heritage Center occupy spare shelf space in the herbarium. Some of them are over a hundred years old. They are not artifacts yet. Fragile as they are, they are still working documents as the staff refines early Rocky Mountain Herbarium plant collection locations for the growing database. These notebooks travel in selected piles from room to room, or to the photocopier downstairs, depending on the errand at hand. Tidily packed cases old and new have to be fumigated for marauding insects, as herbaria have for centuries. Tables are strewn with the shifting geology of student and staff projects that can somehow always yield a spare place to sit. Ernie Nelson—no relation to Aven—the herbarium's curator, a tall man in cowboy boots and pearl-button shirt, is usually somewhere nearby. Watching a student holding a gluey specimen uncertainly over a sheet, he says, "This isn't art." The director, Ron Hartman, is here too, always looking as if he had just come in from the field, wearing shorts in every season. The building itself embraces and organizes the whole enterprise. It is more or less fireproof, one of Nelson's preoccupations in the herbarium's (and any library's) more tentative days.

The connection between plant specimens and books, through the notion of the "library," gives us an entry to understand what specimens are and the kinds of information they might embody, through and beyond botany. An "herbarium" was, originally, a book: about plants, to be sure, an "herbal," but by the eighteenth century also a bound book of specimens.[1] A book that described and contained objects of nature. The modern herbarium is different only in degree: not a book, but an entire library of books, both about and containing natural objects. That the Rocky Mountain Herbarium is in fact housed in a former library, and that Nelson made both tangible realities, gives us a place to sit among the boxes and cases, books and plants, finished and unfinished projects. This will remind us to think about the connections between books and plants, descriptions and objects of nature, remembering that books are objects too, and perhaps seeing what sorts of places we inhabit with both of them.

Imagine we've signed in—you, I understand, tentatively—and cleared a space between a surprisingly vivid Siberian iris (acquired by exchange long ago from the Gray Herbarium for some Rocky Mountain plant or other) and a pile of plants you've never seen before, layered in newsprint, that might have been collected last summer or anytime in the last fifty years. We don't have to catalogue those. What can we do here? I am writing to someone who is reading: we know what books are. These magnificent, intricate things scattered around us on the table can be opened as well.

## EXPLICATION DU TEXTE

The lure of a plant specimen lies in the relationship between what it was in life and what it has become, both a distance and a presence. The specimen has been removed from a place, and transformed by a person who isn't usually available to describe the moment or place of collection; the reality of the plant lingers. Wherever it's found, in its living and pressed forms, a specimen is unique. Pressed, dried, and mounted on paper, it is no longer a living plant, no longer fills three dimensions, and has been removed from the landscape that gave rise to it. Has is it lost too much to be useful or captivating? Unlike illustrations—colorful and realistic as these can be—a mounted plant is in some way the plant itself. It is irrefutable material evidence of life, literally drawn from life. Isolated and framed on its white background, a pressed plant draws attention to itself formally; it invites and allows scrutiny long after its living form would have ceased doing so. The unique and mundane reality of the plant can become secondary to abstract scientific or aesthetic ends. Fundamental to botanical science, taxonomic study locates the plant in a phylogenetic order; aesthetic study responds to the form, texture, and color of a specimen as a composition. (Sculptor David Winter wrote about specimens, "I found that a blade of grass, flattened and isolated on the page, had tremendous graphic power."[2]) Still, the presence of the plant, flattened and dry as it may be, is a silent reminder that there is and was more.

But first, the basics. Living plants become specimens through the process of finding, numbering, pressing, drying, identifying, and finally mounting them on sheets of heavy white paper. Numbering, pressing, and drying plants begin in the field. Identifying and mounting them take place indoors. A plant's collection number from the field follows it into the herbarium, accompanied by an internationally recognized botanical name (with the name of the authority who first identified it), the date and place of collection, and the name of the collector, all presented on a label glued to the lower right-hand corner of the herbarium sheet. The sheet may be stamped with the name of the herbarium as well.

Small plants fit easily on a sheet of newsprint in a press—12" x 18"—but with larger plants, more active arrangement takes place. What the collector has in mind is the size of the standard herbarium sheet: 11.5" x 16.5".[3] Large plants have to be bent "accordion style (V-, N-, or W-shaped, etc.)" when they are pressed so that they fit the finished sheet.[4] When attaching pressed plants to herbarium paper, a collector might be tempted to center each specimen—an aesthetic choice—but some guides caution against it because the pressed pile will become lopsided and unstable.[5] (It is taken for granted that a plant is labeled on the paper "right-side up" as it would be seen in life.) Although scattering plants over the pages keeps the lower end of the pile of specimens from getting thick with roots and stems,[6] most specimens are framed as the central objects of these 11.5" x 16.5" compositions. When the plants are bent to fit the page, the *process* of framing is even more obvious.

Each specimen takes its place with others of its species in a folder, and the folders are stacked with others of the same genus, organized in cabinets throughout the herbarium usually in phylogenetic order. In the Rocky Mountain Herbarium, red folders designate a genus of Wyoming species. This library of plants is a scientific collection, no longer any recognizable plant community. Still, the contemporary work of the Rocky Mountain Herbarium refers to these communities regularly—documenting especially endangered species, those likely to be disturbed by mining, poor range management, or development. A specimen has two "places": one in its habitat, one in a named botanical order.

A specimen in an herbarium also accumulates a variety of meanings by virtue of its place of origin, the habits of its collector, and the cumulative knowledge and disputes of those specialists who study it.

A plant is identified by a name that embeds it in the history and shifting conventions of botanical nomenclature, the ongoing project of naming the order of nature. One plant of the genus *Delphinium* in the Rocky Mountain Herbarium was renamed five times (including different genus names) before becoming a "Delphinium" again, all dutifully recorded in successive layers of plant and botanists' names on the label. A plant's two-part species name, the first of which identifies its genus, links it with other individuals physically resembling it in successively more comprehensive groups: a species, a genus, a family, and a division of plants. Beyond the taxa of the plant kingdom lie all other living creatures. A plant's botanical name links a plant to the classifying habits of the person who named it, much but not all of which is shared by his or her colleagues, either when the plant's novelty is first published, or in the decades or centuries to follow. Needless to say, plant names are not particularly stable. Each one is more like an argument than an object.

The entire system allows botanists to orient themselves in "reading" a plant in many ways. They may look into a comparison with other plants from similar habitats, historical periods of collection, or species, or into the conventions of other collectors and collections. Identify even a single variable in this layered network of relationships around a given specimen—consider, for example, the domain of the inquiring specialist's expertise—and we begin to identify a context in which specimens can be read.

As specimens, pressed plants embody scientific information for botanists. The practice of making and keeping specimens is therefore closely bound up with the history of scientific investigation of natural objects and the conventions of representing them, an amply documented history. An important thread of this history, to which we'll return, hinges on the name, "herbarium."

The "herbarium" was originally a book-length compendium of medicinal knowledge of plants, an "herbal," and many works of this kind were illustrated, commonly so even in the time of Pliny and Dioscorides at the turn of the last millennium. The shortcomings of illustration were noted at the time: "Not only is a picture misleading when the colours are so many," wrote Pliny, "particularly as the aim is to copy Nature, but besides this much imperfection arises from the manifold hazards in the accuracy of copyists."[7] Such remarks suggest Pliny at least valued some form of realism in illustration. Pliny and Dioscorides both observed plants themselves first hand, and recommended field experience to others, a point Agnes Arber and Violet Dickenson make about Renaissance herbalists as well.[8] After centuries of manuscript copying and in a spate of fifteenth-century republication of classical texts, it was clear that illustrations had acquired ends that were not strictly speaking "realistic," and that mimetic accuracy in representation with the growth of interest in the natural world in the Renaissance demanded new technologies for production and reproduction, including learned technologies of perception: what an illustrator expected to *see* in any plant form.

Renewed observation and study of nature was kindled by the availability of classical works (including Pliny and Dioscorides) in print, and the deluge of new flora encountered in European exploration and conquest outside Europe. Realistic illustration aided identification of widely known plants —known by people other than oneself—in the field, and captured images of plants from the "New World," unknown in Europe or the classical world, that viewers or readers might never have seen otherwise. As artists' paintings and drawings became more "accurate" to accommodate curiosity and study of known and unknown plants, printed copies of illustrations, originally accomplished by woodcut, were eventually produced with more sophisticated woodcut engravings and innovations in metal engraving and lithography (and later of course many types of photography)—all media through which an artist's representation of a plant or a more direct image of the plant itself could be reproduced and circulated throughout a community of collectors and scientists. The skill of the artists was paramount in rendering original illustrations both accurate and useful, if botanists were the expected audience; an artist's botanical knowledge (or the botanist's artistic ability) might eliminate one step in the transfer of desirable information from the plant to the botanical community. The skill of the engravers was also important, and again, if an artist was a skilled engraver, or an engraver was especially fluent in translating the techniques of brush and color to etched line and pattern, accurate translation of the original illustration into print reproduction was more likely. Add to this the problem of color—used as part of the original painting or drawing and applied after engraving and printing, if applied at all before color printing was possible—and it should be obvious that botanists and artists striving for naturalistic accuracy in reproduced illustration, from the fifteenth century onward, were fighting against a thicket of things that could go wrong.[9]

This effort toward realistic accuracy in plant illustrations and their print reproduction led to astonishing virtuosity among Renaissance painters and engravers. What was really happening was a shift in the *kind of information* any illustration was supposed to convey, a demand for a new kind of information for which a simpler style would eventually suffice.[10] Rather than a full impression of all the plant's color, texture, and form accessible to paint and ink—beautiful as the result of such an effort can be—botanical illustrations need offer only enough information to identify a plant, information fairly easily conveyed in the stylistically simple drawings that characterize modern (and the most ancient) botanical texts. A botanical illustration does not have to represent the plant itself, but its most visible identifying features—it is a tool, a "map" to the real plant's "territory." What is left of the fuller sensual (even visual) recognition of a plant lies in its elaborate textual description. These simplifications remain preferable to some authors even though photography is available. If you think of the difference between a map of a place and a photograph of it, you will grasp the difference, which hinges on the *use* of illustration. The era of unparalleled aesthetically rich botanical illustration for scientific purposes passed with the sixteenth century.

Changes in plant illustration during and after the Renaissance were accompanied by an emerging scientific orientation toward natural objects, whose first aim was the identification and classification of those objects. Botany emerged as a new practice from the age-old medicinal knowledge of plants, establishing itself as a science related to, but separate from, medicine by Linnaeus's lifetime (1707–1778). Linnaeus understood the value of good illustration—he was especially pleased when botanist, painter, and engraver were all the same person—but he also understood that illustrations alone could never be a perfect record of the information embodied in the plant world. He recommended keeping an herbarium, a collection of pressed plants, as well.[11]

Luca Ghini (c. 1490–1556), an Italian professor holding the first chair of botany established at the university in Bologna in 1534, is credited with keeping the first collection of pressed and mounted plant specimens for study, a practice that spread throughout Europe through Ghini's students. Between the sixteenth and eighteenth centuries, this kind of collection was a *hortus siccus*, *hortus hyemalis*, or *hortus mortus*, before French botanist and physician Pitton de Tournefort used the term "herbarium" to denote such a collection in a publication of 1694.[12] Dry garden, winter garden, dead garden. The herbarium's connection to the garden was not, of course, accidental; physicians had been nurturing gardens for teaching and medical practice for a very long time. Ghini's accomplishment was surely practical; like Dioscorides and Pliny before him, and Linnaeus after him, Ghini would have understood the limited reliability of illustration for plant identification. The plant itself, preserved dry under pressure and mounted on paper, was its own form of illustration, drawn directly from the garden. Gardeners of the *hortus siccus* brought the garden indoors and bound its pages into books, where like all books, it could be read again and again in any season, in any lifetime, barring insects, mold, flood, or fire. One of Ghini's students, Gherardo Cibo, began compiling such a dry garden in 1532 that survives to this day.

Linnaeus dispensed with binding specimens, and stored them loose-leaf, horizontally, so he could rearrange and add to them,[13] an innovation appropriate to a man (and an era) concerned with the order as well as the expanding contents of nature well beyond the confines of the European apothecary's garden. The practice of binding books of specimens (even for multiple copies of published works) persisted into the nineteenth century,[14] but after Linnaeus, the standard practice of plant collection in botanical institutions was loose-leaf, unbinding the *hortus siccus*. The bookishness of the dry garden did not disappear, though. Its transformation into the "herbarium" by Tournefort and Linnaeus gave the scientific plant collection a name borrowed from the old tradition of illustrated medical herbal books.

These books are not merely "outdated," or inaccurate, but possibly charming artifacts. The rules for good illustration were different before the Renaissance. And, as a glance at a modern botanical line drawing of a plant shows, ours is not an era of complex illustration technique, scientifically. What was the purpose of early illustration? In what way may that have survived into the present through the "herbarium"?

Manuscript illustrations were, historically, both decorative and emblematic.[15] Their use as emblems opens another interpretive avenue in the vicinity of our specimens. Manuscript illustrators relied on emblem books and other manuscripts as sources for illustration, copying and recopying emblems for hundreds of years before they were transferred to print early in the Renaissance. This process both magnified the "distance" between the copy and any real plant or animal that may have been the original object of the picture, and created a tradition of stylized images, however poorly drawn, that were icons or symbols rather than attempts at realism. As new plants and animals were found outside Europe, illustrators struggled to portray them on maps and in manuscripts without a known pattern or first-hand familiarity, resulting in pictures of, say, bison and beaver that look utterly otherworldly to us now. (New World plants did not figure as prominently as animals on maps, because they were less formally distinctive from known plants, and therefore less useful than animals as emblems of unfamiliar places.) The stylized quality of these illustrations was bound up with a very old understanding of what images were for; they were "used diagnostically . . . just as banners identified knights."[16] It was the illustration's emblematic use that Renaissance artists abandoned in creating illustrations for new sciences, when illustration became "datum as well as symbol."[17]

At home in their own era, however, copied emblems had widely recognized and longstanding folk, religious, and medical meanings. Herbals were full of images copied by hand, and then copied by woodcutters for printed herbals, "reproducing schematic icons of particular plants."[18] Agnes Arber reproduced a number of these images in her history of herbals, including a fifteenth-century woodcut of plantain that is neatly symmetrical, bearing a scorpion over the central leaf in the composition, and a snake winding among the plant's roots, signifying plantain's use in treating stings and snakebite. She also included several fifteenth- and sixteenth-century images of the mandrake, a plant in the nightshade family with narcotic properties, whose roots were thought to resemble a human being. Each of these illustrations depicts the roots of the plant in explicitly human form; the earliest image, from about 1481, shows a dog leashed to one of the root's legs—a visual reference to the perhaps thousand-year-old advice to avoid digging this plant oneself (tying it to a dog instead, lured by meat just out of reach, to pull the mandrake up).[19]

Though Arber understood how much of this early tradition would be abandoned by scientific botany by the end of the historical period she covers (1470–1670), she attempted to understand the conventions of printed versions of medieval herbals on their own terms. "Some of the figures have a special charm," she wrote, "and, in their decorative effect, recall the plant designs so often used in the middle ages to enrich the borders of illuminated manuscripts." The use of compositional symmetry, lack of realistic proportion, and the inclusion of symbolic elements of a plant's use, meanings, or habitat, were surely intentional, not failures of technique. The artist understood "his work as symbolism," not scientific draftsmanship in the modern sense. "Before an art can be appreciated, its conventions must be accepted. It would be as absurd to quarrel with the illustrations just described . . . as to condemn grand opera because, in real life, men and women do not converse in song." Illustrations "were often so conventional, and the descriptions so inadequate, that it must have been an almost impossible task to arrive at the names of [observed] plants by their aid alone. The idea which suggests itself," she concludes, "is that a knowledge of the actual plants was transmitted by word of mouth, and that, in practice, the herbals were used only as reference books, from which to learn the healing qualities of herbs with whose appearance the reader was already familiar. If this supposition is correct, it perhaps accounts for the very primitive state in which the art of plant description"—as well as illustration—"remained during the earlier period of the botanical renaissance."[20]

Images in herbals were certainly unreliable as representations of the plants themselves, but they were probably "an attempt to preserve the arcane nature of the knowledge" in herbals, "even the very corrupted images of the early printed herbals serving as *aides-mémoires*."[21] That is, if they were not likely used to *identify* plants, it is likely they could be used to *remember* things about them: their properties, meanings or uses, or arcane knowledge of any kind. Identification could take place in person, with someone who knew what he or she was talking about; this is why plant specialists have always understood the value of basic field work among live plants. But the history, lore, uses, habitat, and so on would have to be remembered somehow by the novice or expert—learned and memorized by the novice, and remembered in great quantity over long periods of time by the expert.

If learning and memory were the point of medieval and Renaissance herbal emblems, the more stylized, exaggerated, and unrealistic they were, the more useful in fact they might have been. Medievalist Frances Yates argued that stylized, symbolic imagery was part of a classical rhetorical technique, the "art of memory," renewed and christianized in the Middle Ages. This tradition was carried on by orators and later priests, particularly Dominicans, many of whom practiced and wrote about the memory arts, and were noted for their prodigious memories (including Thomas Aquinas and Giordano Bruno). Preachers, like all orators, have an occupational obligation to memorize a great deal of information in some order; theologians, of the scope of Aquinas, have an entire cosmos (and its order) to work with. For Yates, the longevity of formal memory technique helps explain the character of medieval art, and the persistence of iconic, "unrealistic" emblems and symbolic images through the Renaissance—not in science or technologies of illustration, per se, but through the ongoing practice of the memory arts.

Although Yates's conclusions in 1966 were posed as questions, they are very suggestive questions, especially in light of Arber's similar and likewise tentative insight regarding the quality of herbal illustrations. Here is Yates's suggestion:

> My theme has been the art of memory in relation to the formation of imagery. This inner art which encouraged the use of the imagination as a duty must surely have been a major factor in the evocation of images. Can memory be one possible explanation of the mediaeval love of the grotesque, the idiosyncratic? Are the strange figures to be seen on the pages of manuscripts and in all forms of mediaeval art not so much the revelation of a tortured psychology as evidence that the Middle Ages, when men had to remember, followed classical rules for making memorable images? Is the proliferation of new imagery in the thirteenth and fourteenth centuries related to the renewed emphasis on memory by the scholastics? I have tried to suggest that this is almost certainly the case. That the historian of art of memory cannot avoid Giotto, Dante, and Petrarch is surely evidence of the extreme importance of this subject.[22]

What Yates here refuses to see as evidence of a "tortured psychology" is what Arber and others understand likewise as something other than a "failure" of realism in medieval and Renaissance plant (and animal) illustration.

The art of memory Yates describes originated with the orator Simonides, who wrote a treatise on memory technique; it was demonstrated in a founding story that Yates repeats, noting that many memory treatises began with it. Simonides was invited to dinner at the home of a patron, where the men quarreled over Simonides' pay. Two visitors called for Simonides, and a sudden earthquake destroyed the house, crushing all the people at dinner, while Simonides was safely outside. The bodies were so mangled that no one could identify them. But Simonides demonstrated his technique by identifying the victims according to his memory of the place where each had sat around the table. The visitors outside were the mythic twins, Castor and Pollux, who figure the significance of the quarrel between the orator and his patron: a matter of life and death. With each retelling of this allegory, Simonides (and his followers) displayed a simple form of the technique, and asserted its pecuniary value.

The fuller art of memory works like this: imagine a complex building that you know well, a temple or a mansion (a house will do); if it's a real building, visit it repeatedly. Memorize its rooms so that you can traverse them in any sequence in your mind. Either memorize or imagine permanent objects throughout each room. Once the plan and the contents of the house have been mastered, you can use them to learn and memorize new material, even to see new relationships between objects or ideas you would not have thought about otherwise. Memory figures are often people who are grotesquely, obscenely, or improbably clothed or posed. But memory figures can be objects as well. In either case, memory figures are symbolic of a fragment of the material to be memorized, and can be positioned throughout the house in the order in which they are to be remembered.

Simonides and other sources, even centuries later, described specific "rules for images," and techniques for memorizing both ideas and words. An idea (or word)—*justice*, for example—is to be embodied in an image. Some images could be used repeatedly, over generations—say, a blind female figure holding scales. Whatever the idea, its image should be striking, and perhaps idiosyncratic to one's own memory and experience; images should be placed at regular intervals throughout the house; they should even be placed in "well-lit" areas in the mind. Ideas are figured or refigured from the work of one's own imagination or the storehouse of images already available. Memory for words involves creating an image for every word of a text, a prodigious undertaking that would demand a very large, elaborately furnished house, not to mention a fertile imagination; as an extra step in rote memory, most writers on the subject evidently didn't dwell on it, and neither does Yates. For a list (perhaps a taxonomy), as opposed to an argument or long oration, this form of rote memorization would be very useful. In any case, fluency with fantastical images, their imaginative manipulation and combination, was at the heart of the art of memory. From the classical period forward, the more striking the figures, the more easily recalled they were in the memory mansion. Many of them, Yates believes, found their way into manuscripts and books.

We have access to some of these images still, in the cosmos of astrology and the tarot, including the personification of Justice, which is the figure for the astrological sign Libra, and also appears in standard tarot decks. The astrological sky is almost certainly a memory mansion (one of many), renovated over the centuries to sort human and non-human nature and history into twelve signs and twelve houses, inhabited by symbolically significant, distinctive, and mobile figures—the lively characters of the sun, moon, and planets (of which we now have eight; before the eighteenth century, there were five). The tarot expands this already complex system to include cabbalistic numerology in the four suits of numbered and court cards (forerunners of our playing cards), and adds a layer of figures, the twenty-two trumps, or *trionfi*, which are startling figures in an overarching pageant of human experience, reminiscent of centuries of lists of virtues and vices, as Yates suggests indirectly. Petrarch appears to be a seminal source of contemporary astrology and tarot, as well as medieval and Renaissance memory technique more generally.[23]

Yates explores a variety of other memory mansions, voluptuously imagined to house an enormous quantity of information. Dante's circles of heaven and hell, like many versions of the medieval and Renaissance cosmos (including the astrological sky), were elaborate *structures*, often illustrated, that functioned as repositories of accepted or argued truths and ideas, all vividly figured in vast organized storehouses of memory and knowledge. The structures Yates explores with greatest interest are the "theater" of Giulio Camillo (which was intended to hold all available knowledge, a model of which was actually built in the sixteenth century and which had some resemblance to the architecture of Shakespeare's Globe Theater), and Giordano Bruno's sixteenth-century "memory wheels," which likewise were intended to be comprehensive in capacity (including, presumably, the history of the art of memory itself—a fact embodied in repetitions of the founding story of the art). Both men believed their structures, and the arrangement of knowledge in them, had occult properties, a belief related to heresies for which Bruno would be burned finally in Rome in 1600.

What Yates offers is a compelling case for the importance and widespread practice of the memory arts, including the production of memory figures and the "mansions" in which they lived, from the Middle Ages through the Renaissance. The art would have been familiar to learned people who still lived in an era when comprehensive knowledge could be imagined, when all the books of the world, not just those pertaining to one's "specialty," might conceivably be read. Say, then, that the herbal was illustrated to offer ready-made memory figures for the novice and mementos to the expert—a historical point I will leave others to document; I offer it here as an exercise. If these are memory figures in need of a mansion, what sorts of mansions could these figures inhabit?

The herbal book, as a concrete object, could itself be such a mansion, though it would be more like a hut because the figures would be very crowded. Each leaf of printed paper could be a conceptual "place" if the author or copyist was inclined to organize, and not merely list, the contents.[24] Gardens were a more flexible layout for memory, floral or any other kind. Simon Schama suggests this connection in *Landscape and Memory* (1995). The medieval and Renaissance garden gathered plants and natural forms together with human artifacts (and artifice), organized in space, to teach and remember, as well as delight. In the sixteenth century, Bernard Palissy designed "a garden of 'natural secrets' where adepts and initiates could comprehend the primordial structures of creation," including the four rivers of Eden. Palissy was a hydraulic engineer as well as a naturalist and a chemist; he was also a Platonist, and a comprehensive gatherer of Renaissance understanding. His four rivers, and the four ornamented "grottoes" through which they coursed, no doubt partook heavily of divisions of the cosmos Camillo and Bruno would have recognized. He believed that the "varieties of natural form ought, if correctly discerned, to correspond to the many faces of God. So if the right formulae of inquiry were applied, those laws (and the countenance of Divinity) could be revealed to the learned. It might then be expressed in symbolic, exemplary form. His secret garden was a route to knowledge that was simultaneously scientific and mystical." Botanical gardens were "another way to gather in all the diversity of the natural world, the better to expose its underlying regularity, . . . to re-create the botanical totality of Eden. . . . [A]n exhaustive, living encyclopedia of creation could be assembled that would again testify to the stupendous ingenuity of the Creator."[25]

A garden as a memory mansion, regardless of its original intention or the assumptions about "nature" that gardens as such allegedly embody, expands substantially if we look at it as a structure in which all knowledge might be figured, placed, and remembered (not just knowledge of the nonhuman world). It is a special place, but it may be a mistake to assume it "bring[s] the primal world into the shelter of the garden."[26] This house of memory is not a shelter. It is a structure for organization and recollection. It doesn't just *keep* something (in or out), but displays and demonstrates what is thought to be true both inside and outside its walls. The library is another space like this, conveniently relieved of the burden of being anything but a structure (culturally or historically telling as this structure may be in a given era) for sorting, storing, and making knowledge available. The outdoors itself is another such space—a very big house, the Big Inside we might say, rather than the Big Outside—and likewise not always a shelter.

The *hortus siccus* and its transformation into the herbarium—with its preoccupation with naming the order of the plant world during and after the eighteenth century, and the possibility of adding to and rearranging its contents—embodies both, the endlessly expanding "garden" and library. Both refer to the Big Inside, the living habitat of human and nonhuman beings. The herbarium became botanical nature's mansion of memory. Botanical knowledge is arranged in a remembered order throughout a familiar building (and "building"), from the ferns and mosses through the composites, arranged by division, genus, species, and finally alphabetical order. Botanists are always ready to add a new genus or reduce one to synonymy and rearrange specimens accordingly. The whole shifting complex order is figured by none other than the plants themselves—striking, unique in their endless variety.

## (SE) SOUVENIR

*Eriogonum umbellatum*. Wild buckwheat. Given to a former lover who kept a bouquet of darkened dry buckwheat flowers from an outing with someone else. This buckwheat glowed with the pastel green, pink, and yellow of its living form, under glass in its frame.

*Rosa woodsii*. Wild rose. Sent to a friend because the rose is her flower, her emblem, tattooed on her flank.

*Mertensia ciliata*. Bluebells. For a man to whom all blue flowers are bluebells, while he scours the ground for all kinds of arrowheads.

*Oxytropis campestris*. Loco. A warm inside joke for a therapist friend perches on a shelf in her office.

*Linnaea borealis*. Twinflower. Given to a colleague because of its scientific name, returned to me because of its common name.

Rose-of-Sharon. Lives in the pages of my dictionary, slowly disintegrating, since the last summer I spent with my grandmother.

Anemone. My daughter's middle name. Grows in alpine grassland, as she does.

These are all souvenirs and mementos, though not all of the same kind. They are obviously not scientific specimens either, but they figure knowledge as well as some private sentiment. The memories they embody are complicated, capturing parts of my relationships with specific people as well as my experience with "nature," usually in specific places. Rose-of-Sharon and buckwheat: lost time, places, and people. The quintessential souvenir. The others are not nostalgic. Bluebells: difference in proximity—you see this, I see that, though we are both looking and will enjoy many afternoons like this on this mountain. Roses, loco: recognition over time—I know you, I will see you again. Twinflower: a breach, an impossible knot of interpretation, a grim puzzle. Anemone: where you were born, how you thrive, your long continuous intertwining with me from that day to this. These plants are the specimens that remind me to look for varieties of memory in other specimens.

Specimens inexorably embody memory of all kinds. A specimen soaks it up and exudes memory as an aura. The source of its aura, as Walter Benjamin suggests about works of art, is its uniqueness enmeshed in a tradition. It is flattened but not *flat*; what is left of its three dimensions is a palpable texture of its life in a nonhuman history, and testimony to its history afterwards. Specimens are impossible to reproduce, obviously. The tradition in which they are embedded brings both natural and human history together through memory. Without memory of any kind, "the melancholy man sees the earth revert to a mere state of nature. No breath of prehistory surrounds it: there is no aura."[27] Even a naturalist requires this prehistory—it is her very subject.

An aura is made up of "the associations which, at home in the *mémoire involuntaire*, tend to cluster around the object of a perception."[28] An aura is very definitely a physical effect of a thing (on a person) in a place and its history there, a proximity that can be tasted, smelled, heard, or felt as well as seen. Literally an exhalation, it's not a hidden supernatural light.[29] Referring to Freud's *Beyond the Pleasure Principle* (1920) on the subject of memory, Benjamin's use of the "involuntary memory" signals that this aura is not an effect of consciousness. Objects that concentrate large volumes of involuntary memory (for instance, what I think human or nonhuman nature is, not to mention what I'm doing with this fragment of it in my hand, and what it makes me think about) are deliberately wrested into consciousness as specimens, known and recognized in the scientific order of the herbarium. But we would have to define a specimen as including the paper and the label, not just the plant itself (which has no value as a specimen without the authority of its date and place of collection and the name of the collector), and so specimens cannot escape the aura of a larger memory. With only a pressed plant in hand (or a live one for that matter) and no direction for memory, there is no relationship, no history, no meaning; the highly artificial label actually guarantees the continuity of an aura that extends back beyond the history of science and the scientist. Antisentimental, specimens are explicitly given to (natural) history and the (human) traditions of sampling and describing it; what we would call the personal memory (sentimental or otherwise) is usually lost. That does not mean it was not felt. The scientific specimen both obscures and expresses an aura.

About the work of art, Benjamin says its uniqueness "is inseparable from its being embedded in the fabric of a tradition," a fabric shattered by reproduction. The value and use of the work of art is the result of its being an "original," not a forgery, carrying the authority of authenticity from its beginnings in the hands of the artist through decades or centuries of its history, including "the changes which it may have suffered in physical condition over the years as well as changes in its ownership." What value and use? "We know that the earliest art works originated in the service of ritual—first the magical, then the religious kind. It is significant that the existence of the work of art with reference to its aura is never entirely separated from its ritual function." The connections possible here to indigenous (including European pagan) land-based ritual and memory suggest again how widely "natural objects" might be significant as occasions for nonscientific, but not exclusively personal, cultural memory. The religious cult value of an object "would seem to demand that the work of art remain hidden. Certain statues of gods are accessible only to the priest in the cella; certain Madonnas remain covered nearly all year round; certain sculptures on medieval cathedrals are invisible to the spectator on ground level. With the emancipation of the various art practices from ritual go increasing opportunities for the exhibition of their products"—masses give way to symphonies, frescoes and mosaics give way to paintings. Similarly, unique natural objects and landmarks embedded in ritual give way to objects collected and circulated by scientists in the service of science, more widely visible. Eventually, mechanical reproduction of music in sound recordings, and of the visual arts by photography, bring a depthless art into wide circulation. Likewise, images of "nature" easily possessed in hand replace habitation among nonhuman beings structured and made familiar by revisitation and ritual.[30]

Certain botanical practices suppress the aura of specimens, but they can only do so incompletely. Botany, like all sciences, has been active in the disenchantment of the world, the "death of nature." Let's take modern science at its word, that its revelations are profane. We could say further, following Benjamin, that botanists relentlessly scatter a plant's aura, capturing conscious systematic knowledge from involuntary memory, removing a plant from its traditional place ("nature," literally killing it in the process), and establishing it in a wholly artificial, universally accessible tradition. It's interesting, too, that specimens are *treated as reproductions* even though they can never be reproduced. Botanists regularly collect "duplicates" of a single plant—multiple specimens of the same species from the same date and place of collection—for sale or exchange with other botanists.

But the "type specimen"—one designated as the standard—remains filed at the home herbarium, and herbaria are not identical. Each of these, too, has a history. Herein lie more of many dusty wisps casting an aura around specimens. When we are talking about natural objects as souvenirs of science, it is impossible to erase or reduce their native authenticity, and some part of a person's relationship to them, even if we cannot know the full dimensions of that relationship. Benjamin spares a few tantalizing words on the authenticity of nature (in an essay preoccupied in the end with photography and film): mechanical reproduction of art interferes with "a most sensitive nucleus—namely its authenticity . . . whereas no natural object is vulnerable on that score." Itself endlessly original, nature is the source of all originals, by which I mean the source of all unique and enduring relationships through memory. About the withering of the aura of works of art, Benjamin says:

> The concept of aura which was proposed above with reference to historical objects may be usefully illustrated with reference to the aura of natural ones. We define the aura of the latter as the unique phenomenon of a distance, however close it may be. If, while resting on a summer afternoon, you follow with your eyes a mountain range on the horizon or a branch which casts its shadow over you, you experience the aura of those mountains, of that branch. This image makes it easy to comprehend the social bases of the contemporary decay of the aura. . . . Namely, the de-

sire of contemporary masses to bring things 'closer' spatially and humanly, which is just as ardent as their bent toward overcoming the uniqueness of every reality by accepting its reproduction.[31]

We shouldn't expect Benjamin, urbane denizen for whom Paris was the capital of the nineteenth century, to dwell much on the properties of natural objects. It's enough to say he hit on something essential to nature as an experienced and remembered entity: the relationship is *both* a distance and a connection through perception and memory. An intimate distance. The aura of the mountains and the shadow of the branch ultimately lie in the memory of resting *that* summer afternoon, with a proclivity to follow distant lines and notice shadows. Presumably one would follow and notice again; the objects of perception would vary, but the proclivity to look (and hear, smell, taste, and touch) would remain constant.

Too often we assume the moment of experience and perception is "lost," which is why we collect souvenirs: to remember (or create) what was unique and fleeting, personally. Susan Stewart writes that the souvenir, which might be some natural thing, "must be removed from its context in order to serve as a trace of it, but it must be restored through narrative and/or reverie. What it is restored to is not an 'authentic,' that is, a native context of origin but an imaginary context of origin whose chief subject is a projection of the possessor's childhood." In this way, the souvenir is part and parcel of the decline of the aura Benjamin describes. The souvenir refers to a primordial loss (a modern understanding of both childhood and the past as irretrievable), captured nostalgically, like the full "prelapsarian" world signaled by the representative collection. "We might thus say that all souvenirs are souvenirs of a nature which has been invented by ideology."[32] (Is there any other kind?) Curiously the specimen makes possible both the (public) narrative of a plant's place in the order of nature, as well as reverie on the date and place of its collection by a person. Nostalgia does not explain all collection and recollection.

The form of the collection, which "relies on the box, the cabinet, the cupboard, the seriality of shelves,"[33] itself refers to a longstanding and widespread form of memory, not just those habits of naming, classifying, and ordering what is known that we might judge harshly, not just "scientific," and not nostalgic. Originally, as Celeste Olalquiaga points out, collected natural objects were "read allegorically," "in a universe where worth was measured mainly by objects' ability to stimulate the imagination."[34] These allegories were not personal, and referred not to the past but to the burgeoning present of all available knowledge, which collections of all kinds attempted to pool together. What tied them to memory was not nostalgia (whose contemporary scourge had not been invented yet), but the need to remember what was known. They were (and are) organized spatially in such a way that revealed the accepted meanings of the objects of the collection, which referred directly to human and nonhuman nature, the world at large, as it was—as it was understood to be exactly then.

"Natural collections followed the classification models available to them, mainly the medieval *arbores scientiarum* and *ars memoriae* [tree of knowledge and art of memory], chains of facts whose memorization was considered a form of knowledge."[35] Olalquiaga's dismissal of the memory arts as a tool for organizing information is odd, since she describes the "fragments of an extraordinary narrative . . . constituted in the continuous repetition of anecdotes" that settled around given objects of nature, each repetition adding "a new layer of glamour to something that was already very lightly attached to reality"—in her view, in any case. The aura of natural objects was the result of these layers of repeated stories. And, typically, "naturalia [were] not seen as belonging intrinsically to nature but rather as independent manifestations of cosmic powers."[36] As a result, the "chains of facts" committed to memory actually placed objects of nature in a *particular* order, which connected them, hardly randomly, by the very habit of finding "correspondence or contiguity (similarity or proximity),"[37] to the order of God. As one of Giordano Bruno's characters says in *The Expulsion of the Triumphant Beast* (1584), "*natura est deus in rebus,*" nature is God figured in things.[38] We would do well to remember this outlook. There would be no need to see natural objects as "belonging intrinsically to nature" until about the seventeenth century; human and nonhuman beings belonged to God and each other, in great unifying webs of correspondences and intimacies, resemblances and distinctions, the whole cosmos nested in layers of such resemblances, microcosm to macrocosm. The "prose of the world" was its poetry first; prose would come in huge volume later. And significantly, webs of correspondences would evaporate from science (and scholarly practices eager for scientific "precision") to be concentrated in literature and the arts.[39]

Long since stripped of its most obvious allegories and lore, the order of botanical nature is still organized on the principle of resemblance, forming an open-ended archive of botanical knowledge and material. Plants most like each other in physical form are found near each other in space, and thereafter filed alphabetically. Form determines the place where any species as such might be found—what lies beyond the abstract "place" is the natural place, also recorded.

The long effort to catalogue nature coincided with the long effort to catalogue all knowledge, much of it increasingly available in books, which were, in turn, catalogued themselves. Natural objects no longer bespoke a divine order, books were no longer alive with magic like Prospero's, and structures of organization no longer provided direct communion with God (like Bruno's memory wheels), but knowledge about the things of the world nevertheless moved into old patterns of organizing and remembering their place. Early systems of library classification, like that of sixteenth-century Conrad Gesner, constructed a "tree of knowledge that proceeded by successive divisions," beginning in Gesner's case with twenty-one of them, according to Roger Chartier.[40] Chartier does not tell us what they were, but memory divisions figured by the seven classical liberal arts were common (grammar, rhetoric, dialectic, arithmetic, geometry, music, astronomy—second nature to scholars by Gesner's time, and not an implausible way to suggest organizing a library). Add to these any combination and types of philosophy, virtues, or theological knowledge, in sevens, threes, and fours, and a comprehensive framework resonating with the known medieval cosmos resulting in twenty-one of something is likely. A fourteenth-century fresco depicting the knowledge of Thomas Aquinas has twenty-one divisions (with allegorical personifications as well as historical human representatives of each): the seven liberal arts, along with three theological and four cardinal virtues, and seven other figures that Frances Yates concludes "represent [other] theological disciplines or the theological side of Thomas' learning."[41] Twenty-one divisions would have been a familiar scheme.[42]

Gesner's contemporaries abandoned the divisions and figures of the memory arts in creating systems of library classification, relying on the old technique of alphabetization and whatever list of topics the classifier thought pertinent, creating a flat and abstract order for an emerging understanding of the world itself as a depthless realm of facts. But in defining categories of knowledge, these topics take on additional depth, hardly disrupting the history of memory. A "topic" is a division of memory by place in the traditional memory mansion: "Topics are the 'things' or subject matter of dialectic [rhetoric] which came to be known as *topoi* through the places in which they were stored."[43] Antoine du Verdier imagined an all-inclusive library in the sixteenth century, and published a catalogue that both represented this ideal library, and served as a template by which people might build their own collections. About his list du Verdier noted, "As in the Library divers books are organized, where they are kept as if in their proper place, thus many divers Authors and books are here put *in such an order that at first sight one can locate them in their place and thus remember them* [my emphasis]."[44]

What is striking is not that knowledge (even comprehensive knowledge) was released from the concatenated orders of knowledge passed down through the Middle Ages, but that *books themselves* in a physical, serial order, figured *their own place* in knowledge, the better to remember them, exactly like memory figures. Overt allegory may have disappeared, and along with it the fantastical accretions of memory figures for ideas and things, but the expectation that things were learned and remembered in place and in order did not—from the knowledge of knowledge (in libraries and books) down through all the objects and events of the world (in the herbarium and museum). We might be nostalgic for the lost allegorical world, as Olalquiaga suggests: "after being fragments of a magical universe collected by the scientifico-mystical vision of the Renaissance, naturalia became the scattered debris of that vision, their allegorical impact doubling as the dust of forgetfulness began to cover them with allure of a world—and a whole way of life—now most definitely gone."[45] But its habits of memory were not gone entirely, and show no sign of disappearing. What is striking is that books and other objects, including plant specimens, became their own memory figures without further embellishment.

The library and its microcosm, the herbarium, are a memory of memory, and not just a souvenir. *Se souvenir* was a verb long before *souvenir* was an eighteenth-century noun. Its reflexivity—in English, to recall to oneself, to recollect—suggests the repeated nature of familiar remembered things, not necessarily the neurotic repetition of a lost singularity. The organization especially of a scientific collection reminds us to remember, not just what may have been "lost," but what is believed to *be*. Its very structure teaches us again how to remember: putting things in places, and revisiting them over and over. This is why the story of Simonides belongs at the beginning of a memory treatise. It performs what the treatise is about, a figure for the technique, itself memorable and a reminder to pay Simonides. What the remembered things are and what places they occupy change dramatically over time. But the longevity of the expectation of repeated familiarity with some spatial order is remarkable. One learns by repeated exposure, committing things to memory but also expectation. "We do not need or desire souvenirs of events that are repeatable,"[46] but that does not mean that we don't need or desire to remember these, too.

Moreover, memory of things which have a place in a figurative order becomes very easily memory of places, which is the fund of memory in this tradition. Arranging objects, cataloguing them, ordering them, possibly displaying them (and their order), recapitulates a memory of memory, but also perhaps a placement of place: what or where any "place" is.

In medieval and Renaissance gardens, we might easily anticipate gardener-scholars' "control of nature," and curiosity about the Creation. But further, by creating a place with an order (if perhaps the imagined order of Eden), all creation was grounded in a *specific* place, which was in fact the premise underlying a cosmos ordered by resemblance. A collection in a garden, recapitulating divine order and history, was a reminder that every place was exactly such a place, riddled with signs, bearing the secrets learned people knew how to read, or could learn in a garden, by walking through, paying attention, and remembering. Perhaps the garden is a mnemonic of what a place is, not what "nature" is. This suggestion has little to do with gardens or "nature" as such, and more to do with what a place might be good for, including but exceeding the objects of nature. We should not be surprised to find evidence of this phenomenon outside the garden.

Unpacking his library, Benjamin invites us to look in on him and his books, "in the disorder of crates that have just been wrenched open, the air saturated with the dust of wood," "not yet touched by the mild boredom of order." The aura of his books saturates this essay and opens his memory. "Everything remembered and thought, everything conscious, becomes the pedestal, the frame, the base, the lock of [the collector's] property. The period, the region, the craftsmanship, the former ownership—for a true collector the whole background of an item adds up to a magic encyclopedia whose quintessence is the fate of his object." There is an elegiac, "magical side" to the collector: "As he holds [his objects] in his hands, he seems to be seeing through them to their distant past as though inspired." This sounds like simple nostalgia. But Benjamin is more interested in the "childlike element" of the collector, who renews the world (and creates himself) by acquiring new (old) things. Memory has to be ordered somehow; presenting the ways he acquires his books "is something entirely arbitrary[,] . . . merely a dam against the spring tide of memories which surges toward any collector as he contemplates his possessions." Tellingly, travel is among Benjamin's techniques in finding books: on "the wide highway of book acquisition," visiting various shops, "how many cities have revealed themselves to me in the marches I undertook in the pursuit of books!"[47]

These places belong to the fate of his books because they belong to the history of the collector. Having filled nearly all his cases between noon and midnight (one might say working through memory places, symbolically significant divisions of the day), Benjamin says, "Other thoughts fill me than the ones I am talking about—not thoughts but images, memories. Memories of the cities in which I found so many things," he begins, leading us all over Europe to "the rooms where these books had been housed" when he was a student, "and finally my boyhood room, the former location of only four or five of the several thousand volumes that are piled up around me." The past is present at this late hour, palpable, continuous. Amid the smell of wood dust, late at night he says, "I put my hands on two volumes bound in faded boards," scrapbooks his mother had made for him, "the seeds of a collection." It is from this origin, in a family, from remembered places, with books Benjamin can touch and smell, that the essay draws rapidly to a close. The continuous renewal of a book's fate does not mean in the end that books come (falsely) alive in the collector; rather, he says, "it is he who lives in them. So I have erected one of his dwellings, with books as the building stones, before you, and now he is going to disappear inside, as is only fitting."[48] We don't see the exact order of the collection, but we definitely see the force of memory bound to places he knows, a memory mansion of books from places that defines the history and memory of the collector.

In another instance, Benjamin clearly demonstrates that this propensity to spatial memory collects more than nostalgia. "One-Way Street" is a self-generating tour through found objects along a route, "named Asja Lacis Street after her who as an engineer cut it through the author."[49] We begin at a filling station, and proceed through a variety of signs, each illuminated by aphorisms, observations, speculation, forward and backward in time, some themselves describing this structure of memory and thinking along a route: "The power of a country road is different when one is walking along it from when one is flying over it in an airplane. In the same way, the power of a text is different when it is read from when it is copied out. . . . Only he who walks the road on foot learns the power it commands. . . . Only the copied text thus commands the soul of him who is occupied with it, whereas the mere reader [and the air traveler] never discovers the new aspects of his inner self that are opened by the text [and the road]."[50] This observation is "filed" under Chinese Curios, anticipating Borges's Chinese encyclopedia which this street and its haphazard (and sometimes searingly funny) accumulation of things and even lists resembles. "Commanded" by the one-way Asja Lacis Street and signs along his route, all recreated in his hands, Benjamin considers *both* his "inner self" opened by these objects as well as the outer world full of crowds, politics, instincts, and intellectual history, a great amalgam of impressions. He is both copying and thinking, walking and creating a landscape, remembering and learning.

"One-Way Street" is a serious daydream, imaginatively structured as a place by flight of both imagination and memory. Another serious daydreamer, Gaston Bachelard, lays before us a cornucopia of places—specifically houses and parts of houses—in which memory and experience curl, huddle, and sprawl. His aim in *The Poetics of Space* (1958) is to "show that the house is one of the greatest powers of integration for the thoughts, memories and dreams of mankind." Simonides would have agreed. Bachelard repeatedly sides with the daydream over the world of realistic fact, inviting the reader over and over, on the threshold of some space, to enter and dream: "we have to induce in the reader a state of suspended reading. For it is not until his eyes have left the page that recollections of my room can become a threshold of oneirism [deliberate dreaming] for him." Bachelard disappears into his room as Benjamin disappears into his book building. It would be a mistake to go in *looking for them*—a truly nostalgic project; they warn us anyway that they've disappeared. What we should find rather are our own routes, our own rooms. Some but not all of this daydreaming is nostalgic. Like Benjamin's return to routes, "a dreamer of houses sees them everywhere, and anything can act as a germ to set him dreaming about them." Bachelard's "houses" include landscapes and shapes in them; bringing in nests, shells, trees, Bachelard uses the house as the dream portal to his "album of concrete metaphysics."[51]

These are the last words of a grand tour through intimate spaces, and it is important that they come at the end rather than the beginning. Had they opened his book, we would perhaps expect a comprehensive, fixed catalogue. As it is, they conclude a thought on "becoming," which, through one of Rilke's images of a tree, has "countless forms, countless leaves," which "in spite of everything, illustrate the permanence of being." *If*, he says, "if I could ever succeed in grouping together all the images of being . . . Rilke's tree would open an important chapter in my album of concrete metaphysics." Bachelard's humility is rhetorical, like his repeated observations that realists will dismiss his daydreaming. The "album," introduced at the very end of a primer on being and becoming through spaces, is neither possible nor desirable. It is the final invitation to dream and think in which Bachelard climbs his tree and vanishes.

Bachelard's formulaic dismissal of the dismissive critic anticipates a reader's claustrophobia in all those cabinets and corners, just as we might be nonplused by Benjamin's (re)collection. The atmosphere may be cloying; maybe he needs to get out more. "When I relive dynamically the road that 'climbed' the hill, I am quite sure that the road itself had muscles, or rather, counter-muscles. In my room in Paris, it is a good exercise for me to think of the road this way. As I write this page, I feel freed of my duty to take a walk: I am sure of having gone out of my house." Really? Or, "Unfortunately, being, as I am, a philosopher who plies his trade at home, I haven't the advantage of actually seeing the works of the miniaturists of the Middle Ages, which was the great age of solitary patience. But I can well imagine this patience, which brings peace to one's fingers." Impatient with Bachelard, we would in fact be speaking of ourselves. After all, "The house we were born in is more than an embodiment of home, it is also an embodiment of dreams. . . . Our habits of a particular daydream were acquired there."[52] If we reject the "house" where he was born and the habits it engraved in him, we are by rejecting it perhaps thinking of our own.

There are many routes to, through, and about places, and at the same time about memory, which are taxonomies, lists, catalogues, and collections damp or dusted over with the concerns and memories of their collectors. Many of these lie outside, or, like the herbarium, come from outdoors, and not surprisingly they say as much about us (and dreaming) as they do about "nature." Simon Schama, who offers us many gardens and landscapes, begins *Landscape and Memory* on the Thames of his childhood, finding a flow of history in spite of himself, densely mapped. He follows it through wood, water, rock, and finally all three together—to a specimen, in fact, the "wild hairy huckleberry," a prize of Thoreau's backyard, and of our very long walk with Schama. "[T]he backyard I have walked through—*sauntered* through, Thoreau might exclaim—is the garden of the Western landscape imagination: the little fertile space in which our culture has envisioned its woods, waters, and rocks, and where the wildest of myths have insinuated themselves into the lie of our land,"[53] a serious pun. Through the huckleberries in Thoreau's hand, Schama closes his book with an observation: "For this is what the unappetizing little fruit, finally, had to tell Thoreau, and us: 'It is in vain to dream of a wildness distant from ourselves. There is none such. It is the bog in our brain and bowels, the primitive vigor of Nature in us, that inspires that dream.'"[54]

This surely is the primordial history and enduring connection between memory and place. Our relationship to both courses through the objects that embody memory of places, tells us what places are, have been, or could be, how they are related to one another and to us, and who we are and have been as well. Some of this information is scientific—the scientific collection does not escape the relationship between memory and place—but the scope of the relationship is broader than what science has become. That we make (our) nature through and in place, through and in objects, including natural ones, is a point that goes further than arguing the social construction of nature.

The association of memory, traversed space, encountered objects, and created places, various enough in European traditions, is not even exclusively European in practice. Working from another direction, David Abram describes the Australian aboriginal experience of landscape through Dreamtime songs and stories: "it is the land itself that is the most potent reminder of these teachings, since each feature in the landscape activates the memory of a particular story or cluster of stories," and he adds, "even within European culture there is a celebrated example of this propensity, albeit in thoroughly altered form," the art of memory. Using Frances Yates, Abram notes that "the classical orators had to construct and move through such topological matrices in their private imaginations," while "the native peoples of Australia found themselves corporeally immersed in just such a linguistic-topological field, walking through a material landscape."[55] Abram also refers to the work of Keith Basso, which documents Western Apache memory and knowledge residing in revisited landscapes with eloquent place names in *Wisdom Sits in Places* (1996). For Western Apache people, simply speaking a place name, or saying names in sequence—without any further gloss in conversation—invokes a whole library of useful, amusing, or otherwise immediately necessary information. Textual literacy profoundly altered the techniques of reading and remembering the places of the world, inspiring catalogues, orders, collections of all kinds, and books themselves, including memory treatises. But the phenomenon of intimate distance—the experience of knowledge, memory, forms of encounter, hearing, reading, and reciting—remains bound with objects in places, both abstract mnemonics and real landscapes. Some of these effects are public and shared, like the most explicit use of the herbarium specimen, the Dreamtime landscape of cultural memory, or the murmured but continuous landscape memories of Europe. Some are private and idiosyncratic, beyond the name and date of a collection to an experience in a place that is not ours and probably not recorded anyway. All of us nevertheless learn and remember what we believe is ours from the Big Inside.

### ROUTES

Olalquiaga enters and leaves the "artificial kingdom" of marine kitsch through her glazed hermit crab, Rodney. Benjamin steps out into Marseilles, Moscow, Berlin, Paris, and streets of the imagination. For all its closed interiors, Bachelard tours a vast poetic imagination through the objects and forms of the house. Susan Stewart, interestingly, critical of nostalgic forms, assembles a still catalogue of the perversions of memory as close and airless as Bachelard's cellar. We see her neither come nor go, leaving only an inscription over the museum door, "For my mother and grandmothers." What she gives them we can guess is bound with warm familiarity, however complex; what she gives us are artifacts she's left behind, scrupulously documented (one might say scientifically), tagged specimens, though not when or how she collected them: a collection without a collector and without order: not a place. Because the book "is a collection and not a chronicle,"[56] no narrative or reverie can transform the collection nostalgically. She has secured for herself the authority of a collection that escapes nostalgia because it is not "hers"; Stewart lays out a taxonomy, not a route. It's not clear where we go if we follow her. Those who give us routes give us landmarks, invitations, and significant divisions of both "outside" and "inside." They do not say "you are here," but something more like, "this is one way to get anywhere." Remember a river, unpack your books, read poems and dream of houses, look into the glassy eyes of a captured crab (he will look back).

Or in this case: sit in this old library, the herbarium, with these plants, and remember. Remember where you learned to read the prose and poetry of the world. Remember how you learned to remember. The plants on the table, whose origin in nature is both convoluted and certain, for which a full and precise memory will leave only traces on the labels, are not a place to finish but to start. Where would you go from here?

Tarot trump card, XIV, "Art," from the Crowley-Harris deck. Reprinted with permission of the Ordo Templi Orientis.

ROCKY MOUNTAIN HERBARIUM

364912

UNIVERSITY OF WYOMING, LARAMIE

GEN ROCKY MOUNTAIN HERBARIUM
PLANTS OF WYOMING

GENTIANELLA DETONSA (Rottb.) G. Don
ssp. ELEGANS (A. Nels.) J. Gillett

Carbon County T 14 N, R 85 W, Sec. 18
Sierra Madre; Cow Creek, ca 10.3 air mi WSW
of Encampment.
Local in bog.

10 August 1985 9,300 feet

Roger L. Williams 171

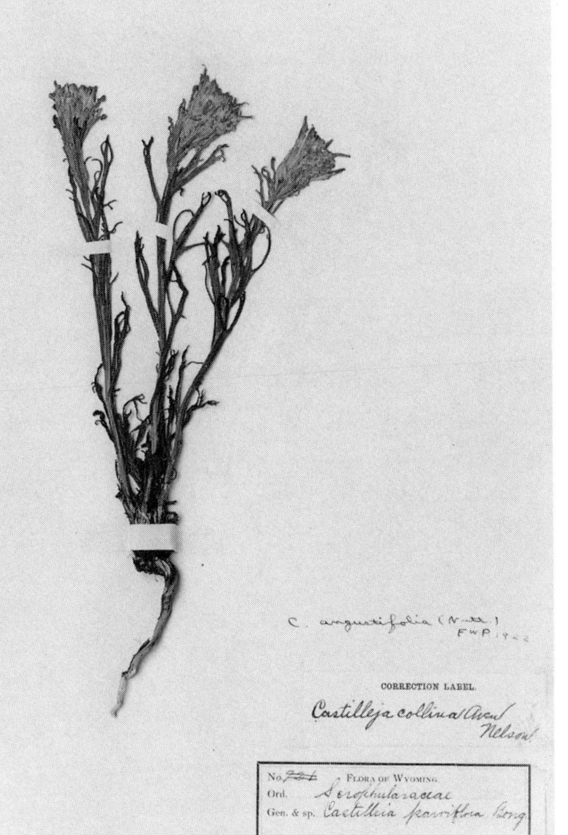
C. angustifolia (Nutt.)
FWP 1922
CORRECTION LABEL.
Castilleja collina Aven Nelson
No.
FLORA OF WYOMING
Ord. Scrophulariaceae
Gen. & sp. Castilleia parviflora Bong.
Loc. Bear Creek
Coll. B. C. Buffum Date 7/21/92
HERBARIUM OF THE UNIVERSITY OF WYOMING.

ROCKY MOUNTAIN HERBARIUM
721
UNIVERSITY OF WYOMING

C. ochroleuca A. Nels.

C. angustifolia (Nutt.)

FWP 1942

CORRECTION LABEL.

*Castilleja collina Aven Nelson!*

No. ~~[illegible]~~ **FLORA OF WYOMING.** 120

*Castilleia parviflora Bong.*

*Pole Creek. Among sage brush*

Coll. Aven Nelson. *June 2.* 1894.

Herb. University of Wyoming.

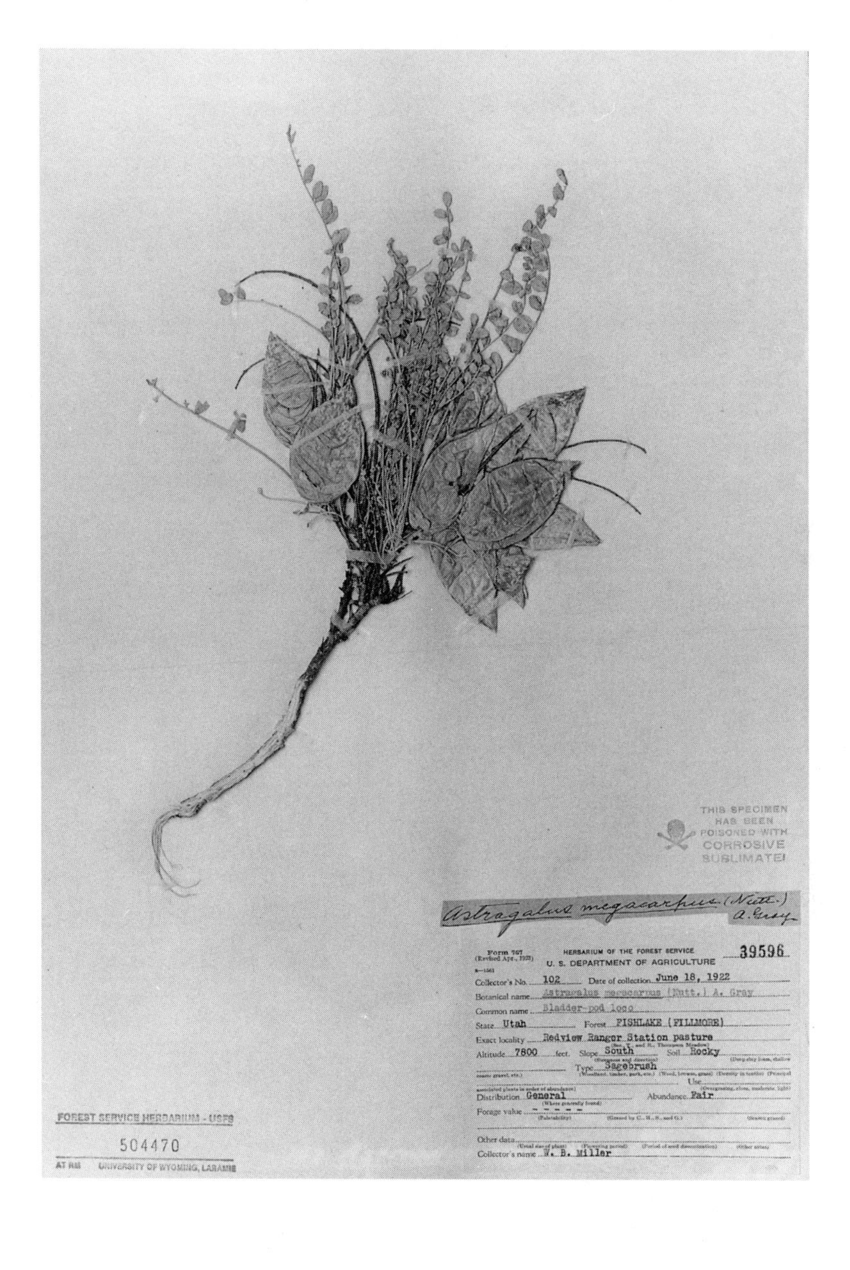
THIS SPECIMEN
HAS BEEN
POISONED WITH
CORROSIVE
SUBLIMATE!
Astragalus megacarpus (Nutt.) A. Gray
Form 767
HERBARIUM OF THE FOREST SERVICE
U. S. DEPARTMENT OF AGRICULTURE
39596
Collector's No. 102
Date of collection June 18, 1922
Botanical name Astragalus megacarpus (Nutt.) A. Gray
Common name Bladder-pod loco
State Utah
Forest FISHLAKE (FILLMORE)
Exact locality Redview Ranger Station pasture
Altitude 7800 feet.
Slope South
Soil Rocky
Type Sagebrush
Use
Distribution General
Abundance Fair
Forage value - - - - -
Other data
Collector's name W. B. Miller
FOREST SERVICE HERBARIUM - USFS
504470
AT RM
UNIVERSITY OF WYOMING, LARAMIE

Geranium viscosissimum
Fragaria
Potentilla arguta
Eriogonum umbellatum
Potentilla quinquefolia
Lithospermum ruderale
Castilleja sulphurea
Foeniculum
Two mustards
Snowshoe hare
Valeriana acutiloba
Senecio integerrimus
Frasera speciosa
Lupinus caespitosus
Antennaria microphylla
Astralagus agrestis?
Serviceberry
Creeping Oregon grape
Prairie smoke
Balsamroot
Sticky geranium
Meadow rue
Antelope bitterbrush
Calypso bulbosa
Clematis occidentalis
Grouse whortleberry
Silvery lupine

Lonicera utahensis
Streptopus amplexifolius
Gooseberries
Currants
Honeysuckle
Paradise
Fritillaria atropurpurea
Aquilegia caerulea
Lonicera involucrata
Black elderberry
Western thimbleberry
Western serviceberry
Cascade mountain ash
Highbush huckleberry
Red raspberry
Ribes
Engelmann spruce
Douglas fir
Lodgepole pine, Pinus ubiquitus
Bracted lousewort
Showy gentian
Mountain bluebells
Viola canadensis
Valerian
Spring beauties
Mahonia repens

# Album

ARIES

Ruth Elizabeth Ashton was born 29 November 1896 in Roxbury, Massachusetts.

Ruth told Aven's biographer, Roger Williams, that she was "reared on Martha's Vineyard where her interest in plants first blossomed during her childhood."[1]

Former student Jane Ramsey, who took classes from Ruth in Rocky Mountain National Park and interviewed her in 1984, said she was "devoted to botany" and that knowing plants was a "life-long passion."[2]

Friends described how she "came alive in the midst of flowers, wild or domesticated."[3]

TAURUS

Ruth's parents, Willard and Grace Ashton, were evidently both high-minded and fairly wealthy. They ran a settlement house in Boston, and began looking into recreational and real estate opportunities in Estes Park, Colorado, in 1905.[4] Willard commissioned a design by Frank Lloyd Wright for an inn in nearby Horseshoe Park, but discarded the design and built an Adirondack-style lodge instead.[5]

As a young woman, Ruth had sufficient resources to buy 240 acres above Estes Park herself in 1925, and named her property Skyland Ranch.[6]

5-25-56 *written a few miles up canyon of Lake Fork of the Gunnison where Sue has just caught one nice fish in very muddy & high water—saw an ouzel about which flew up side canyon where is a small stream of clear water . . .*[7]

GEMINI

Aven Nelson hired Ruth in the herbarium in 1930 because her work on a popular guide to Rocky Mountain National Park plants "squared well with his own ambitions for a dual-purpose manual."[8] Once she began work, Aven described her "as very competent and as sympathetic to the proposed format of the revised manual."[9]

In a Christmas card in 1931, Ruth's friend, Anna Lute, wrote, "I had no inkling of your plans. I did think however that you seemed happier than usual the evening of your nice party—and I am sure with your many common interests your life together will be the happy one which we all wish you."[10]

James Feucht, editor of *Green Thumb* magazine, thanked Ruth in 1961 for her series on gardening with Colorado wild plants: "We are always pleased to receive your well-written and authoritative articles."[11]

Ruth was well informed, gentle as a teacher, "humble with her knowledge." Jane Ramsey remembered how Ruth taught a class about fragile tundra ecology in part by teaching students how to walk through it, from one rock to the next.[12]

CANCER

Grace Ashton suffered an illness that precipitated the family's first move to Martha's Vineyard early in Ruth's childhood. After 1905, Grace went off with the children for months at a time to Iowa, Illinois, Estes Park, and Martha's Vineyard, interrupting their formal educations but teaching them herself at home. She believed they could learn more from the outdoors than they could in a classroom.[13]

Grace was interested in botany and loved flowers; with her mother's copy of Gray's *Manual*, Ruth collected flowers and leaves around Estes Park and identified them for her mother. For Grace's fortieth birthday in 1905, Ruth and her sister decorated the cake with forty different species of wildflowers.[14]

Ruth returned to Colorado after college to work. She remained in Colorado after her husband's death.

Friends remembered Ruth as a "'shy, retiring and very private person,' not easy to know."[15]

Illustrator and friend Beatrice (Bettie) Willard said of Ruth that she had "a deep affection for, devotion to, and comprehension of the land and its diverse plant cover. . . . She expresses profound quiet reverence for the natural world through everything she says and does."[16]

Orra Phelps (1895–1986) was another woman whose mother's intense (in this case professionally trained) interest in plants, and equally intense family instability, gave her a medium for both enjoying and learning about the natural world, and creating her own world of work and companionship all of a piece with the Adirondack wilderness where she lived. Her mother botanized to be alone; Orra's botany and Adirondack mountaineering was social.[17]

*Reflections—objects passing behind me (as I sit in the car) are imperfectly reflected in the opposite glass of the windows—by turning to reality I see the true, correct image.*[18]

LEO

Ruth intended to run a girls' camp at Skyland Ranch. This never materialized, but as a young woman she taught informally in Rocky Mountain National Park, hoping to become a park naturalist. Decades later, after her husband's death, she was able to teach there regularly as a naturalist.[19]

She published five field guides over the course of her career, illustrated guides to places she visited often herself. Her favorite was the 1969 *Handbook of Rocky Mountain Plants*. She thought the *Plants of Zion National Park* (1976) was her most beautiful.[20]

At Skyland Ranch, at the age of ninety, the "wildflowers in the meadow, many transplanted by Ruth, delighted her, as they always had."[21]

She bequeathed most of her property to close friends who enrolled it in a permanent land trust.[22]

VIRGO

Ruth arranged her husband's final residences for his comfort and pleasure, moving with him to Oklahoma for a few years in the 1940s to escape the long winters of Wyoming and Colorado. Moving back to Colorado in 1949, she planted a garden he could enjoy. She moved him into a nursing home when she could no longer care for him herself, keeping a house nearby, reminding his daughters to write to him, and old friends to visit.[23]

In her 1961 article series on gardening with Colorado "wildings," Ruth provided a list of useful species, and mentioned where these were commercially available; some nurserymen, like Harry Swift of Boulder, were happily making more Colorado specimens available all the time, but her recommendations were horticultural, not commercial. Presumably those interested would find and relocate their own plants. She had seen fendlerbush on limestone cliffs at Mesa Verde State Park, and she had had one for several years; "We hope this species will soon be available from nurseries." About pussytoes, she wrote: "I collect it in the wild . . ."; creeping juniper was still difficult to obtain from dealers, but she wrote, "I have collected several plants which are thriving."[24]

*Trail Ridge 6-9-59*
*On top at 9 A.M. with Bettie. Windy but not very cold at Rock Cabins. Primula ang. abundant and showy in bloom,—big clumps and scattered small groups & singles—Eritrichium, fairly numerous phlox just beginning, geum, mertensia, draba beginning to bloom, no silene nor Arenaria—Rock cut—nothing out Tundra leaves: Ranunc. adonis abundant & beautiful—leaves inconspicuous, Caltha coming, sedum in fl bud, Besseya in bloom in the Cobrezia, Thlaspi well out, Floydia in fl bud, Saxifrage rhomb. fl bud. Low willow around little ponds in bloom, both staminate & pistillate the pistils well out, stamens just beginning.*[25]

LIBRA

Ruth married Aven Nelson on her birthday in 1931, and began a long partnership traveling and collecting and working with him in the Rocky Mountain Herbarium.

She traveled often with women friends and collaborators. Two of them completed the illustrations for field guides: Beatrice Willard and Dorothy Leake. In her seventies, working in Zion National Park in Utah, she traveled with her illustrator, Tom Blaue.

*May 21. Packed up—took pictures of B. Fremontii & Mt. Peale—mist came over the mountains & a few rain drops. Beth & I dug seedlings of the Maho-*

*nia (B. Fremontii) repens was there also. Sun came out—2 lovely humming-birds came to the yellow fls. There were orioles, summer warblers etc. in the cottonwoods. Left about 10.30—Decided to skip Natural Bridges as weather was very threatening & we drove thru a heavy shower. Found Peraphyllum soon after leaving Monticello—lunched in a juniper forest & dug a few rooted layers of Peraphyllum. Near Dove Creek the red plowed fields, young green wheat & sagebrush made lovely color picture against the dark sky—saw lots of beautiful shrubs of Peraphyllum . . .*[26]

SCORPIO

Ruth could not get work in Rocky Mountain National Park as a naturalist in the 1920s, even though she attended the Yosemite School of Natural History in California to improve her credentials; she believed being a woman was the problem. The park did not hire a female naturalist until 1955.[27]

Some of her friends believed Ruth was "stifled by her marriage to Aven. Only after she was widowed, for example, did she resume long visits to her beloved Skyland Ranch which Aven had found boring because it lacked a large variety of plants."[28]

SAGITTARIUS

Ruth finished high school in a girls' boarding school on Cape Cod, and began college at Mount Holyoke. She completed some of her college credits at the University of Wisconsin in 1924. Though she took all the botanical courses available at Mount Holyoke, she graduated with a major in English. A family she'd met in Madison hired her to work at their camp in Colorado, near Longs Peak (and near Estes Park).[29]

She began graduate work in botany at Colorado Agricultural College in 1925, completing a flora of Rocky Mountain National Park for her thesis, which was published in 1933.

"She'd throw her hands back and kind of clasp them behind her hips. She had about a three-quarter list and it was a pretty good angle for scouting ten feet in front of her. That's the way she walked everywhere. She was in that pose about 100 percent of her waking hours and it served her quite well."[30]

Ruth's property above Estes Park sweeps down a hillside from the pine forest. The view pours into the cupped valley below, and up thousands of feet onto the face of Longs Peak and its close companions.

*Black Canyon of the Gunnison N. Monument*
*June 15, 1958*
*. . . Climbed up the so. side of pass—found many fine large clumps of Eritrichium in full bloom—also Trifolium dasyphyllum, Thlaspi, potentilla and the finest tufts of Androsace subumbellata that I have ever seen. Photographed the Eritrichium and would have taken the Androsace & Thlaspi but the wind was blowing a gale. . . .*

*In Gunnison we found out from a F.[orest] S.[ervice] man that the road over Black Mesa from Cimarron, to Crawford & then on to the Nat. Mon. was open & in good shape—so serviced the car there & left with the intention of camping at Soap Creek—turned off at Sapinero & drove about ten miles—decided we'd missed the way to Soap Creek so camped near Pioneer Point—a lovely high mesa, but below Black Mesa—had good aspen wood to burn and sagebrush perfume. I wakened at 4.45 and watched the sun light come—first on the rim of Black Mesa, then on the tops of the tallest trees, then on the shoulder of the mesa north of us—As it came down the green wall of Black Mesa small groves of white stemmed aspen in grassy glades came into clear focus—The sky was a great wide clear dome, supported by the low rim of faraway mesas (encircling us but far off, except for the wall). We started rolling at 8.45, down the hill & then up the mesa side in long switchbacks—stopping every little ways to look at flowers.*[31]

CAPRICORN

Working on her thesis, she asked Aven Nelson for help identifying plants; that was how they met. The Nelsons' earliest collecting notebooks show her still learning plant species. She wrote in the genus name of a plant, if she knew it, and left the species designation to her husband. By 1940 the collection notebooks were almost entirely in her handwriting.

The Albuquerque *Journal* "published a perfectly correct story under a rather snickering headline: 'Savant, 72, and Bride, 35, Expect to Work and Play Together, They Say Here.' A press photograph also revealed the great disparity in their ages and conveyed the impression of an old goat nibbling tender grass."[32]

Tom Blaue, working with Ruth as a young man, remembered her as teacherly rather than motherly, and demanding. "She went to bed every night thinking about what she was going to do the next day. There wasn't too much casual about her in the sense that she never, to use the modern terminology, 'went with the flow of things.' She had plans and agendas and she adhered to them quite closely."[33]

*Def. of* species
*Dr. Stabler—Aiken—5/8/56*
*"A species is a group of organisms which look alike, which produce offspring which look like themselves, whose members are able to interbreed and produce fertile offspring." In a species there is always variation and the great problem is* where *to draw the line. "Subspecies have a meaning" but don't overlook the* species *name. See R.J. Peterson on subspecies in "A Field Guide to Western Birds."*[34]

*May 29, 1938*
#2461 Solidago squarrosa *Muhl. A rare goldenrod*
#2462 Solidago flericaulis *L. another rarity*
[She took three specimens of each.][35]

AQUARIUS

When Ruth wrote botanical keys for her field guides, her "original keys were remarkably simple; but to achieve that simplicity, she omitted keys to a number of plant families, genera, and even species that she treated in her text. It may be that the casual user of the book was never inconvenienced by the omissions—or never discovered them." Williams revised the keys in her *Handbook of Rocky Mountain Plants* to include "not only . . . all groups and species mentioned in the text, but . . . numerous additional species likely to be found within our range that cannot be treated in the limited space of an introductory text."[36] Williams's revision "completes" Ruth's scheme by filling in "omitted" keys and adding more of his own.

Williams does for the field guide what anyone might do with the documents of Ruth's life, filling in blanks with a purpose defined outside the subject. No biography (or scholarly project of any kind—even a field guide) avoids this condition. Faced with inevitably incomplete data, what happens is a relationship between two subjects: the one making inquiries, and the stuff followed as if it were a trail left behind. The changes Williams made in the book underscore the amateur practice of taxonomic botany—Williams's avocation. Any other person approaching her work or any of her documents would do the same thing, filling in a picture of some kind by whatever habits —patterns—of thinking and knowing or even being they already have.

What her original text offers among the lively physical descriptions (and keys), like many other popular field guides (which often do not provide keys), is what we might call "lore":

> BITTERROOT, *Lewisia rediviva*, an amazing plant brought back from Montana by the Lewis and Clark Expedition in 1806, revived after many months in a dried condition and, when planted, bloomed. It was given the name rediviva, meaning to live again. The new genus was named in honor of Meriwether Lewis, leader of the expedition. The little green tufts of slender leaves come up in winter on bare ground or under snow, but they have withered almost completely away by blooming time. The showy many-petaled pink blossoms dot gravelly terraces and stony places that receive abundant moisture in early spring. The thick roots were a staple food for the Indians of the northwest, who boiled them until the bitterness disappeared. The plant, the state flower of Montana, is known to grow in only a few places in Colorado and southern Wyoming, but it is abundant farther north and west, and in the Grand Canyon area.[37]

About silver buffaloberry, she writes that its "female flowers produce quantities of oval, red or orange berries, which are sour but edible."[38] About *Pedicularis*, she notes that the "name *lousewort* is an old English plant name, and *Pedicularis* is merely the Latin form of it. It was once believed that cows who ate this plant became infested with lice."[39] Elk thistle "is the species that saved Truman Everts, a lost explorer in Yellowstone in 1870, from starvation. The peeled root and stems are pleasantly flavored and nourishing, and they are eaten by elk and bears."[40] *Erigeron coulteri* was named "for John Merle Coulter of the University of Chicago, who, as a young man, had accompanied F. V. Hayden on exploring expeditions and collected plants at high altitudes, in Colorado in particular. He prepared the first manual of Rocky Mountain botany, which was later revised and expanded by Aven Nelson."[41] (This is not the only time her husband's name appears in her text.)

Scientific classification systems did away with exactly this sort of embroidering centuries ago. Her attention to habitat, historical associations, ethnobotany, common and Latin names, as well as the taste of edible things, and a tendency to remember her husband's contributions, give us footholds outdoors and in the book, something to remember a plant by or learn a little botanical Latin—threads linking plants with other things we might be interested in, or linking them with her own memory. Beyond the habitat and physical appearance of plants, there is no taxonomy organizing whatever might appear in these entries. An older natural history, they are rather layers of fragments around each object.

PISCES

The Ashtons separated in 1905, during the first summer the family spent in Estes Park.

Ruth was widowed 31 March 1952. In April she was on the road collecting; in July, vivid prose description breaks abruptly into the serial collection list for the first time:

*7/1/52 from Highland Camp (alt 10,008 ft)—O'Dea, Bakersville—7 miles west from top of Loveland Pass—3 above Silver Plume steep road 3 miles to Stevens Mine, about 12,000 ft. (above timberline) much Rydbergia in full bloom, Erysimum nivale some in seed—Silene acaulis at its best—Columbine (coerulea) coming into bloom—lovely buds—good color—some A. saximontana, phlox condensata (?), Claytonia meg. in bloom, penstemon Harbourii one plant in bloom a few others in bud in scree above old trail leading up eastward from buildings, Senecio soldanella in bud, lovely Mertensia—erigerons (see collecting notes). Ranunculus eximus in swales recently freed from snow —the mountains about very patchy with snow—much pinus aristata along upper part of road & apparently in an open stand on very steep wall to east of valley—cones at high tips dark purply blue—Floydia abundant in edge of forest just below t[imber] line.*[42]

7-7-59 *Leaving Skyland—Gaillardias coming into bloom, Galium boreale in full bloom on lower Long Hill—Pent. alpinus, wild roses out, miner's candle, pink geranium, potentillas, seeds of pulsatilla gone—Penstemon semilatter-alis making a great show of color—susans in bud, harebells coming out.*[43]

Ruth Ashton Nelson died 4 July 1987.

Balsamroot

Fritillaria atropurpurea

Fragaria glauca

Phacelia sericea

Frasera speciosa

Grass

Sedge

Willow

Snow

Gentiana dentata

Halogeton glomeratus

Rumex venosus

Artemisia tridentata

Artemisia frigida

Rosa woodsii

Oxytropis deflexa var. sericea

Epilobium canum

Mayflies

Gunfire

Antennaria microphylla

Phacelia sericea

Populus tremuloides

Artemisia tridentata

Lupinus argentus

Phlox longifolia

Phlox multiflora

Atriplex or Sarcobatus?

Eriogonum umbellatum

Antennaria parvifolia

Green Castilleja

Senecio

Penstemon

Pink

Medusa head

Stanleya pinnata

Penstemon

Long-spine shrub with red flappers (saltbush)

Linum perenne

Cleome serrulata

Tiny sage

Scarlet globemallow

Dune daisy

Prickly pear

Allium

Yellow ball-head, eriogonum?

Red-paddle seed bush

Phlox longifolia finished

Atriplex

Sarcobatus vermiculatus

Chrysothamnus viscidiflorus

Verbena bracteata

Atriplex canescens

Chrysothamnus nauseosus

To:
*Ruth Ashton Nelson*
*c/o The Roseate, Blossom-filled Fields on the Light-kissed Hill-crests of the New Jerusalem*

# Letters

12 September 2000

Dear Mrs. Nelson,

I'm writing because I have been working with the papers you left to the University of Wyoming archive, and your collection books at the Rocky Mountain Herbarium. I have been interested in the work of the herbarium since I came to the university in 1997 and have been grateful to be able to look at your collection lists and those of your husband. I met Dr. Roger Williams soon after I read his biography of Dr. Nelson. He and his book have been very helpful in forming a general picture of Dr. Nelson's life and some of your collaboration with him, but reading through your papers has raised a number of questions that I would like to ask you directly, if I could. I am happy to answer any questions you may have about the project I am working on. I understand Dr. Williams was concerned that I not in any way jeopardize the work of the herbarium, and I want to assure you that my interest in that institution and the work you and Dr. Nelson did to sustain it stems from a deep respect for your careers here.

I am especially interested in what drew you to botany, and to Colorado, as well as the nature of your collaboration with Dr. Nelson.

By the way, I have very much enjoyed using your *Handbook of Rocky Mountain Plants*; it has helped me learn a great deal about the plants here around Laramie.

Thank you for your consideration.

Sincerely,
Frieda Knobloch
Assistant Professor
American Studies
University of Wyoming

13 October 2000

Dear Mrs. Nelson,

I came across some things that were both interesting and startling, and you are the only person I can ask about them. When I started looking at Dr. Nelson's collection books, I was looking at where he collected, along with anything he noted about each plant; I didn't pay much attention to the collection numbers themselves. (Since I am not a botanist, the scientific names have been a preoccupying challenge but I use your books to translate for me.)

Recording Dr. Nelson's collections in New Mexico, I noticed that in 1931 he began his serial numbers again with #1—immediately after your marriage. And the new entries are headed "Collections of Southwestern Trip 1931—Aven & Ruth Nelson." Most of what follows remains in Dr. Nelson's handwriting. But in your USDA seed collection that year, I noticed that you remarked on a few plants as likely transplants for your garden. This was such a departure from Dr. Nelson's usual style of recording plants that I wanted to ask you: did you talk about how you would share the list, and what kinds of information were to go into the collection list? Did it seem natural to you to bring your own habits of observation into a list you were then sharing with your husband?

I was curious about something else too. When you were listing plants on your own, very often Dr. Nelson would complete a species name where you had given a genus, or you left a space where he wrote in the name of the plant later; sometimes he corrected your names. Did the fact that Dr. Nelson was your husband affect the tone of this education in any way?

I was delighted to see so much of your handwriting in the lists after 1931; I did not realize how much of Dr. Nelson's later collecting was shared with you.

Sincerely,
Frieda Knobloch

1 January 2001

Dear Mrs. Nelson,

I've been reading about women in botany, women working in the field with husbands. There's a wonderful book by Cora Steyermark, *Behind the Scenes* (1984), describing the many years she collected with her husband, Julian, for the Missouri Botanical Garden while he completed his flora of Missouri—perhaps you knew them. She wasn't trained as a botanist, and her book is the only writing like it I've found, describing the gear they carried, the landscapes they visited, many plants and animals they saw and their pleasure pointing things out to each other, the people they encountered, all in the course of finding, pressing and drying plants. She condensed thirty years of her observations and experiences into a single collecting season, March through October, a sensible way to sort things out. You would recognize her attention to the passage of time in the seasons of plants and animals she observes. I'm sure you would also recognize the long time driving from one place to another.

The book gave me some idea what it might have been like for you and Dr. Nelson to collect and travel together. Referring to her husband and herself as X.Y. and X.X., the biological world (and "work") is closely intertwined with their companionship. I wonder, if you had written a book like this, how you would have chosen to represent the close tie between work and companionship in your life with Dr. Nelson. This kind of material is very hard to come by because it's not just about botany or a career, but the lived experience of a marriage, much of which happens to take place outdoors. I have wished for some time that you or Dr. Nelson had written more about things like this, what you saw together, how it felt to work this way, really how it felt to *live* this way, through botany.

Unlike Mrs. Steyermark, you were eventually a full partner in the field, I know. Well, more than that. The collection books are almost all in your handwriting throughout the 1940s. Dr. Williams didn't know that—he didn't have the lists among the papers he used for his biography. It doesn't surprise me really that you shared Dr. Nelson's collecting, but I wonder how you thought about your own work with him, if you both understood how much of it was yours.

Your travel journals describe whole landscapes and the people you were traveling with, and remind me very much of how Cora Steyermark wrote about collecting with Julian. You describe beautiful places, colors, little events of everyday travel, striking views. It is obvious you were enjoying what you were doing, and if botany was the point, there was still much more going on. But there aren't many of these journals—loose papers really, dating only from after Dr. Nelson's death—and I began to wonder if you had other journals. I don't know if you only began writing this way in the 1950s, or if for some reason you kept other journals out of the archive here at UW or the library at Rocky Mountain National Park. One of your collection books suggests your descriptive writing emerged just after Dr. Nelson died: in your notebook for the summer of 1952, you took the time to write out a long description of a day collecting in Colorado, which was surprising after reading years' worth of simple lists. Where did this come from? Maybe you had just misplaced your journal on that trip. I know I am prying, but this seemed important.

I'm torn between wishing there was more material in your own hand to draw from, and simply recognizing you brought a different sensibility to writing and collecting than Dr. Nelson did—you were willing to describe things aesthetically, try your hand at written description. Your field guides would depend on it. But I'm afraid of burying you in a history, even a history of your own work. (I have even less interest in rewriting a history of your husband's work.)

Most of the questions I have about you I can't answer. You and Dr. Nelson surely saw much more than specimens. What did you see in each other along the way? Did your work together change you? Was all that, too, woven into the collection lists somehow like the bouquet you were given on your first trip to Europe? How should I read them?

Sincerely,
Frieda

13 February 2001

Dear Ruth,

I feel like I'm reading a code, or a poem. (You were a botanist, but you were also an English major; maybe you will understand what I mean.) Here are the things you left for others to read, the biggest volume of them published guides to various places, as if to say, "look this way." Away from you? Or are you visible there, in the gentle style and rigorous keys? And I've read through thousands of entries in the collection books that say in their own way, "look here." I'm trying. The collection lists are both intimate and abstract as a record of your relationship with botany and your husband, time actually spent outdoors. Were some plants more important to you than others? I mean symbolically. John Muir shared ferns with a woman he loved, the way other couples have a song. Your bouquet and your garden transplants, the new series beginning with #1 when you married, these are the most obvious signs of your lives rising through the surface of a botanical list, but I'm sure there are others. I don't have enough information to read for more.

Frieda

7 March 2001

Dear Ruth,

My mother gave me a biography about an Adirondack botanist you may have run into at Mount Holyoke—Orra Phelps. She reminds me of you, making a career of something she learned to love as a girl. Like you, she traveled and botanized (and did her Adirondack climbing and guiding) with many women friends. She reminded my mother of her mother (a lifelong nature lover cut from the same cloth as you were, and Orra Phelps, and countless other people), a composite really, since Phelps was a medical doctor too, as my father's stepmother was, trained in the 1920s. There are hundreds of questions to ask about you, but I wonder how many of them are really about you after all. The Phelps book was important to me all on its own as a gift—my mother has heard a lot about you, and your husband, and knows I've been busy with these lives and my own.

I'm thinking now about women's education in the twentieth century, too. Along with marriage. And botany. Histories of these things. Most of what's written about women who were botanists is a nineteenth-century story. You remind me of Kate Brandegee, maybe. Both she and her husband were botanists. But what does that mean, that she reminds me of you? Brandegee wasn't a popular writer. Cora Steyermark accompanied her husband, like Edith Clements. (I wonder what you would have thought about Edith, typing her husband's minute plant community observations from one narrow margin to another, filling page after page with her work, typing, along with his, observing. There's something maniacal about the way these pages look, like a list gone mad.) Maybe Anna Comstock comes closer as a parallel, in insect studies anyway, since she definitely worked with her husband, collecting and illustrating, and made a whole career for herself in nature study, eventually heading the department of Nature Study at Cornell—but she was more than a generation older than you, and an academic. Somehow I can't picture you in a permanent academic appointment. Brandegee was a botanist like your husband was, collecting and publishing plants. I don't know, really, that she was anything like you.

I hate to put it bluntly, but Brandegee and Comstock are famous, and you aren't. You weren't a "major achiever"—a phrase from a historian of science looking into married women's accomplishments as botanists; there were a number of women like you who pursued their careers in part through marrying botanists, but the "major achievers" were widows, or never married. (Is this a surprise?) I can't bring myself to ask why you didn't do more. Why *did* you marry? Why this man? To make sense of you I have to compare you to somebody (my grandmother doesn't count); to write about you as a botanist I have to compare you to other botanists, or other scientists anyway, but this doesn't make enough sense to me. You achieved something. I need your help.

Frieda

20 April 2001

Dear Ruth,

I had a nightmare last night, after putting your travel journals aside and trying to sleep. It was about "peer review." I was facing a large committee, most of them women. It felt like a trial. The conversation I suppose has been rattling around in my head a long time, bits of people's comments to me, things I've read.

*Are you absolutely certain they shared a field notebook?*

Yes.

*There is no question that the handwriting is hers?*

No. He shared notebooks earlier with other assistants, too.

*That is very unusual.*

Maybe he didn't know better.

*What were their ecological views?*

They were botanists, not ecologists.

*In the 1920s and 1930s and 1940s? And they knew Edith and Frederic Clements?*

Yes.

*And there is no discussion of ecology in their work?*

Not in their written work. They identified habitats, but not how these develop or change, or respond to development. In their teaching . . .

*You have little record of that.*

I have enough to know they taught reverence for and understanding of natural landscapes in the field.

*But no text.*

They were typical nature lovers, like many other people at the time.

*Like many scientists, too, better known than these two. His degree was not prestigious.*

That's true.

*One might say illegitimate.*

One might say.

*She graduated from Mount Holyoke College. You are aware of the importance of the long lineage of female botanists and naturalists from Mount Holyoke College?*

She took many botany courses at Mount Holyoke.

*She majored in botany?*

In English.

*She was not a major achiever.*

Not in the sense you mean.

*Even her colleague Weber in Colorado did not think much of her as a botanist.*

That's what Williams told me, I don't know that I believe that.

*I came across Nelson years ago when I was reading the papers of Margaret Ferguson, a botanist at Wellesley College. Nelson wrote asking her to nominate him for the presidency of the Botanical Society of America. I thought it rather odd and pathetic he had to ask.*

Perhaps.

*They made no new interpretations, no theoretical advances?*

No.

*Their interest seems very limited.*

Ruth, did you ever read Virginia Woolf? She wrote about the Angel in the House who nagged her to flatter and praise the man whose book she was reviewing; she had to strangle this apparition and throw the inkpot at her repeatedly or she would never have written a word. There are other angels now, just as formidable, but they have tenure.

I'm going outside for the summer, I'm taking your book, I'm taking your field notes and your husband's. I'm taking my daughter with me, to Alaska at least. (I'm only sorry I can't take the dog too.) Most of the time I'll be in Wyoming, many places you would have seen and known well. I made a small press. I bought some pencils and one notebook for drawing, another for a journal. I don't think it's possible to find you, but if I don't look at some living things soon I'll go crazy. I'll write when I get back. One thing I regret: that my grandmother didn't live to travel out here with me. She would have admired you, and would have loved "following" you. I'd be able to show her so much that was beautiful and new to her.

Frieda

12 September 2001

Dear Ruth,

Somehow I doubt this will surprise you, but what happened this summer is as obvious as that bouquet in your "field notes" from Europe. Maybe if your papers had been more complete this wouldn't have happened. On one hand, this is unconscionable. Unprofessional, at least. But on the other hand it's true—it's more true than whatever story I could tell about you achieving this or that.

You may be a specimen of something but anymore even specimens aren't what I thought they were. When I was looking for plants, everywhere associations tumbled through them. As I put things in the press, one look at them brought whole days back to me—did this happen for you? I think it must have—people and things I was thinking about. There's a scarlet gilia that captures everything I remember about the day I drove out of Grand Teton National Park, through Hoback Canyon, in the snow, tired, pleased to see these slender red trumpets everywhere, now that I knew what they were. Like meeting friends. A cut-leaf erigeron I pulled out of the gravel on Barber Lake Road up in the Snowies reminds me of someone I know, since he scattered his mother's ashes in the woods along that road, and skis effortlessly down it in the winter. The first things I collected—marsh marigolds and glacier lilies—I was with my dog, who bounded off into the snow towards them, licked cold snow-melt from icy scoops in the tundra where the marigolds were—but also found the fallen and broken eggs of a nest, bear tracks, many small animals to sniff out, her energy and circling back collecting me the essence of dogness. She was good company. I finally decided all these things weren't distractions. The effort to *exclude* them became the distraction. I can't know what you saw, or felt, or remembered, but each of the plants I collected is a looking glass to step through. The memories and associations are vivid even months later.

I learned a fair share of botany, too, though—in fact this other mysterious thing I stumbled into would not have happened if I hadn't been looking so intently at plants, reading and rereading your keys. I don't know how or when it happened, but at some point I could look at something I'd never seen before and guess its family, sometimes even its genus, and there it was, a name. The plants are what they are with or without a "real" name, but whole families appear—lines of comparison, maps of resemblance—where before there were only so many different *things*. A tiny *Rumex* growing spindly in the roadside in Yellowstone is blown up to huge proportions in the begonia dock in the Red Desert—big flowers like paper puzzle boxes. Or the fine serrated leaves, spread like a palm on one *Potentilla*, are stretched out along a stem, pinnate, for another. Mustard is mustard whether it's white, yellow, or bright purple, two inches tall or three feet. Whether the flowers are complicated by dozens of little heads and long leaves only at the base, like death camas, or the leaves climb up the stems and the flowers hang singly, like twisted-stalk, both of these plants are lilies. A pistil ending in three parts, flowers in six parts, and long pointed leaves.

I wished you'd told me more about the penstemons. I didn't have a very good magnifying glass. I have to confess I gave up on the paintbrushes, too. I admire your stamina knowing more about them than whether they're red or salmon-colored. This was a surprise—how little color tells you. So many flower guides are arranged by color, without keys. Maybe you can find something quickly, if it's common and showy enough to be in the book, but then you miss too much about form. What I remember about the plants I identified with your keys was a patient process, repeated, holding things you must have touched and smelled and pulled apart, and looked at very closely.

I kept thinking I was missing something when I "forgot" I was in the Tetons—how could anyone not look up at those famous ridgelines? But I was pretty absorbed in the monument plants, a tiny *Draba*. In your travel journals, you wrote about the vistas you saw, but also pistils and stamens, minute and palpable. An impossible range of things you felt or remembered or did could be condensed in a single view of Black Mesa, or a flower gone to seed on your own hillside. There is plenty to see and remember at every scale. Plenty to *do* really. I kept thinking *this is how learning works*. Reading forms. Shifting from one scale to another, one thought to another, allowing experience to just happen, paying attention.

I was listening and looking for everything you and your husband did not write down, not just botany (I know this was a hopeless project)—I could have taken a class for the botany, but that's not what I was after. All I knew was that you enjoyed this work and worked together. I already doubted it was all about botany. Learning and looking pulls memory and association in, and adds layer after layer of memory in turn, far beyond the official professional intellectual part. Going back to a place, or recognizing a plant you know, intensifies this effect. That's what I enjoyed about it anyway, how remarkably a single plant allowed legions of thoughts and memories and feelings to alight, like butterflies carrying pollen from one powdery place to another. Mixing things up. Sorting things out. Germinating a whole life through this work. So much the better to have had companions in it—how lucky you were.

Frieda

27 November 2001

Dear Ruth,

I told you I'd answer your questions about this project, but I have many of my own, some of which have become pressing. Why I insisted on writing about you and Aven, for example. "I want to write a biography of this couple's field work," I'd try to explain; how field work brought work and play together, companionship, knowledge of the living world, through and beyond botany. But why you? "You're trying to get blood from a turnip," a friend said. The simple reason was because you lived and worked here, outdoors much of the time, together, and—obviously—I'm envious. Maybe I'm in the wrong field. Or the wrong life. Do you know how rare it is that anyone ever says a word about why they study what they study? I don't mean the big professional pronouncements about why some subject or other is important, but if they *like* it, what place it has in their own lives. More than that, how living *takes place*, with work, learning and reflecting. Somewhere definite. People don't say the simplest things because, I figure, they are not simple.

It wasn't just botany that drew me to you. You told Janet Robertson you'd had an "eventful" childhood. I know something about that, a family disruption, how flowers and plants can come to mean something particular in important places with important people. I knew that a long time before I knew any real botany. I wonder if you went outside for the same reasons I did, getting away, but also bringing flowers and leaves to your mother, reconnecting with her. Orra Phelps did the same thing. My grandmother was that person for me. Even in your most desultory notes your appreciative eye reminds me of my grandmother's. You made me think that "environmental" awareness of any kind might have a source that quiet, that personal, that deeply set in the occasions we have for learning anything.

You also had things I wanted. What struck me about your work and your husband's wasn't how "important" it was but how whole it seemed, which may be important in a different way. Of course I couldn't have really begun to see that if I hadn't learned a little botanical Latin and gone outside. I needed to go outside here. You made a life here. You had a partner here. The work you did brought things together. This is not a scholarly interest, really, but then again it is: where any scholarship comes from has to have roots in what people remember from a larger life, what they fear, or love, on a very fine scale. What millions of biographies get written in intellectual rebus? Even the fact that a book or essay is published is a milestone in someone's effort to keep a job, a roof over their head. Sometimes, that's the only significance I can see.

I can't write a biography. I can't even write a narrative—there's not enough to work with. An analysis won't say anything new about botany, women's or anyone else's. I can't abandon you, though, because following you and Aven reminded me of too many things (people really) that are important to me. How I learned anything, ever. I kept running into all this, like ghosts, all summer. And you taught me something in spite of an incredible distance between what you knew and what I know—how is that possible? But that's been true of every teacher I had. I know you can't write back to me. Still you gave me a way to read this landscape, my work in it, the many routes in which all the rest of my life pours into it. I don't know exactly what I can give back.

Frieda

4 June 2002

Dear Ruth,

I have to finish this long reckoning, but afterwards I know what I want to give back. A book about the Red Desert. I think you would understand why.

Thank you, Ruth.

Sincerely yours,
Frieda

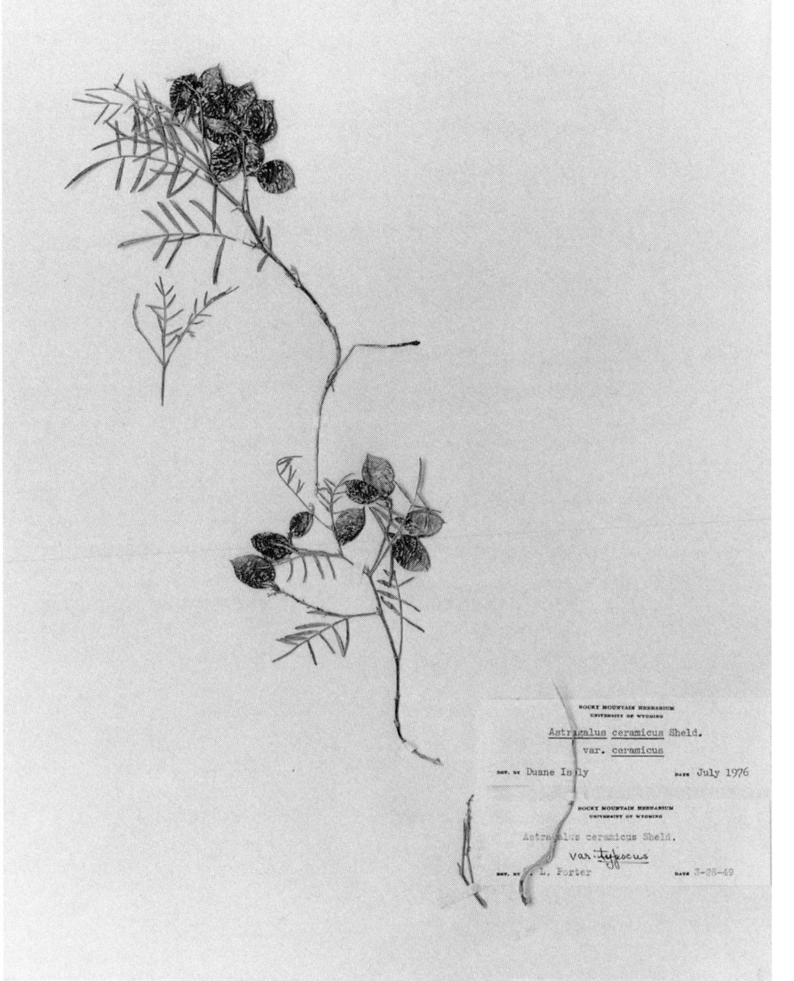


Accession No. 14133
RECEIVED FROM
U. S. Nat. Herbarium
1899
HERB. UNIVERSITY OF WYOMING.

From
THE UNITED STATES NATIONAL HERBARIUM.
PLANTS OF UTAH.
Astragalus pictus Gray.
Kanab. Altitude, 5500 feet.
No 5286 ac MARCUS E. JONES, Collector. May 22.

Chrysothamnus parryi
Artemisia cana
Chrysothamnus viscidiflorus
Delphinium nuttallii
Lupinus argentus
Atriplex confertifolia?
Chenopod
Russian thistle
Spider
Oenothera nuttallii
Oxytropis besseyi var.
obnapiformis
Astralagus purshii
Oenothera caespitosa
Atriplex concertifolia
Artemisia tridentata
Chrysothamnus viscidiflorus
Eriogonum umbellatum
Lupinus argentus
Senecio
Agropyron cristatum
Tragopogon dubius
Antennaria
Iliamna rivularis
Sedum lanceolatum
Shrubby potentilla
Purple erigeron
Gilia
White geranium
Few strangers
Sticky geranium
Dandelions
Bluebells
Snow
Fireweed
White spirea
Oatmeal
Utah honeysuckle, red
berries
Parasailers
Eriogonum umbellatum
Balsamorhiza sagittata
Campanula
Gilia
Agoseris
Hawksbeard
Splashy paintbrush
Mint
Senecio gone poof
Yellow composites
Achillea millefolium
Cicuta douglasii
Heracleum lanatum
Thimbleberry
Spreading dogbane

Physician art thou?—one, all eyes,
Philosopher!–a fingering slave,
One that would peep and botanize
Upon his mother's grave?
. . .
Shut close the door; press down the latch;
Sleep in thy intellectual crust;
Nor lose ten tickings of thy watch
Near this unprofitable dust.
—William Wordsworth, "A Poet's Epitaph"

# Habeas Corpus

On 15 July 1959 in Glacier National Park near Avalanche Creek, Ruth Nelson wrote enigmatically in her travel journal: "Reflections—objects passing behind me (as I sit in the car) are imperfectly reflected in the opposite glass of the windows—by turning to reality I see the true, correct image."[1] If she saw the true, correct image when she turned away from reflections, what was most true and correct was that she had turned and looked for herself.

She spent a great deal of her life looking directly for things. As a young man, Tom Blaue remembered working with Ruth on her *Plants of Zion National Park* (1976, still in print), which he illustrated. She "rarely, if ever, missed anything. She assumed a certain posture when she was looking for flowers, 'which was all the time,'" Blaue told Janet Robertson. "'She'd throw her hands back and kind of clasp them behind her hips. She had about a three-quarter list and it was a pretty good angle for scouting ten feet in front of her. That's the way she walked everywhere. She was in that pose about 100 percent of her waking hours and it served her quite well.'" Catching Blaue looking at a bird in a blazing Utah sky on a hot day, she snapped uncharacteristically and said, "Tom, you'd get more done if you'd keep your eyes on the ground."[2]

Ruth may not have missed much, but *looking for her* is an exercise in blanks. She gave hours and boxes of her husband's papers to Roger Williams for his biography of Aven. She left a little correspondence, photographs, a few article typescripts, and her collection lists. She left slim fragments of travel journals from 1958 and 1959. She herself of course is dead.

And Alice? A devoted nineteenth-century wife, mother of Aven Nelson's children, helpmeet in the field, companion, housekeeper, and hostess. Her diary is not available. The Nelson family lies under the turf in Greenhill Cemetery in Laramie: Aven, Alice, Neva, and Helen. Ruth is elsewhere.

They are all scattered like ashes.

We couldn't find Ruth (or Alice) if we tried, though there is enough information to understand that if we wanted to, we could fit her into a history of "gender and botany," "women and science," "wives of scientists," which would begin with something like this in mind:

*Taxonomy I:*
Scientists
Botanists
Female
Married
Collaborated with spouse
Varieties:

1. facilitates husband's professional work
2. independently active (more likely if she completes education or begins research or publication before marriage)
3. delays until, or resumes independent work after, separation from or death of husband

*The diagnosis of an individual may show considerable variability of form over the course of her lifetime, including all three described varieties. Subvarieties are common. Exclusionary conditions limiting population of the taxa: generally race, class, and gender; in botany, gender to a significantly lesser extent, depending largely on marital status, education, and historical period (cultural factors). Habitat also widely variable, some indication that U.S. western and less prestigious institutions were historically more suitable for employment of married women botanists.*[3]

*Taxonomy II:*
Scientists
Married couples
Varieties:

1. "Peaks of Collaborative Success: The Nobelist Couples"
2. "Couples Beginning in Student-Instructor Relationships"
3. "A Spectrum of Mutually Supportive Couples"
4. "Couples Devolving from Creative Potential to Dissonance"[4]

*Each taxon includes women among the population, and assumes collaboration as fundamental to the practice of scientific work; women's negotiations with the "second shift" of housework and childraising may be assessed; variations of experience over time and across disciplines may be compared. Habitat undifferentiated.*[5]

Depending on the scholarly sieve, we could also describe Ruth Nelson within the social history of westerners, community life of Laramie or Estes Park, or the lives and outlooks of field guide writers. Or, depending on your cynicism, cast all the Nelsons off as unremarkable white middle-class people of a certain generation, may they rest in peace, etc. All histories in which their purpose, ultimately, would be to disappear. I can't claim that Ruth Nelson won't disappear here, too, but it is her life that makes other things visible, fragments of her experience that call up questions, memory, and insight.

I notice these particular people because I bring something of my own to them. A recognition—even a recognized distance—enlivens a relationship, a call and response in which there are cues about being a person in a world that includes human and nonhuman beings, the matrix of living and nonliving things. Looking in the direction of Ruth Nelson, I'm not looking for the body—the ideal body of evidence that adds up to this or that historical object, or the body of scholarship—or for what made Ruth Nelson "important." I'm looking at an ordinary woman looking, forming connections with other people and the living world. Recognizing that I cannot look directly at her, a very complicated amalgam of her and me looks back. I am looking at and experiencing a relationship.

The point would be to understand some of the texture of relationship, including but exceeding relationship with "nature," through a practice of relationship. Specifically, empathy. Therapist Judith Jordan writes, "Empathy is central to an understanding of that aspect of the self which involves weness, transcendence of the separate, disconnected self. It is, in fact, the process through which one's experienced sense of basic connection and similarity to other humans is established." A process of mirroring and differentiation, empathy has an "'as-if,' trying-out quality to the experience, whereby one places oneself in the other's shoes or looks through the other's eyes."[6] Elsewhere Jordan writes that in the process of mutual empathy, "Growth occurs because as I stretch to match or understand your experience, something new is acknowledged or grows in me . . . I accommodate to your experience and therefore am changed by our interaction. I am *touched* by your experience." Empathy is "not a static mirroring process, but an expansive growth process . . . new aspects of self are expressed and each provides that opportunity for the other. This is growth through relationship."[7]

Jordan is writing about affective growth, an intrapsychic development, and the role of empathy (as both mirroring and differentiation, which has both emotional and cognitive components) in the process of therapy itself as well as within the remembered experience of her clients. And of course this relationship pertains between two people in direct communication. But some form of empathy is possible with fragments left by another person—not perhaps with that person's experience directly, but indirectly, and with the fragments themselves. Art historian James Elkins describes the process of being physically moved by empathy in the presence of images, feeling pain, size, as well as a range of emotions, and notes that "the reaction is at its strongest when I am looking at bodies, probably because the origin of empathy is in the body."[8] Creating analogies is the cognitive, intellectual form of this elemental capacity for empathy—analogies capture incomplete but compelling resemblances between things; we inhabit them, in thought, to see something both partially known and new.[9] What is "true" about the outcome is that it is a trace of a relationship.

The structure of empathy that Jordan describes, and that Elkins explores in response to images, is the structure of growth and learning Gregory Bateson described as central to all biological and cognitive life in *Mind and Nature* (1979). Bateson once asked students to pretend they had never seen a crab before, looking at the body of one he had brought to class, and attempt to describe how they knew this object was once alive. Obviously the students could not really achieve a mental state in which they had never seen a crab before, but the exercise invited them, through an object, to deliberately consider *and enact* what a living form is in response to a body unlike their own. The students' description hinged on a kind of empathy: what they recognized in the crab were qualities of aliveness they could recognize because they themselves were alive, formal qualities discerned by the formal, patterned quality of perception and (re)cognition.[10] The "pattern which connects" mind and nature (and unites Bateson's book) is this very process of recognition and differentiation, both enacting and perceiving similarity and difference with another being, a relationship characterized by perception, partial recognition, and learning.

People, images, and objects are all occasions on which relationship can be formed, shaping patterns of knowledge, affect, and identity. A person is born in a context of adult care (or more likely some combination of care and carelessness), and develops in networks of connections beyond home and family. The broadest ground of relationship is the world itself, human and nonhuman, including those sets of things and beings we think of as nature and society. (Relationships that pertain "out there" in turn affect the particular context of any individual's "beginning.") Individuals develop in relation with other beings and things both close to and far beyond home.

Steven Holmes explores this understanding of the self in relation to the environment in his book, *The Young John Muir: An Environmental Biography* (1999). Building on object relations theory borrowed from psychotherapy, Holmes sketches out the significance of "the environment" in individual psychological development—the relationships forged between a person and the physical environment, as well as important relationships with other people, as the fullest context in which individual development can be understood. He cites Clare Cooper Marcus, whose *House as a Mirror of Self: Exploring the Deeper Meaning of Home* (Berkeley: Conari, 1995) argued that psychologists' models of development rarely consider people's relationships in the context of specific physical environments, and that other specialists interested in environmental relationships rarely look into the emotional significance of specific environments in the development of individual people.

Writing a biography, Holmes is keenly aware of the need to make sense of an individual at a fine scale of analysis that must, for him, include emotional response. Writers like Yi-fu Tuan discuss "larger groups—whole societies or cultures—rather than the lives of particular individuals," and even though Tuan borrowed the word *Topophilia* from Bachelard, Tuan's *Topophilia* isn't about *philia* at all but perception, attitude, and worldview. Ecopsychologists like Theodore Roszak describe "humans' deep connection to the planet . . . as an environmental ethic or worldview," which Holmes knows is not useful in his "more descriptive work of analyzing those specific connections that are operative (and in fact created by) concrete lives in particular historical and natural contexts."[11]

Object relations theory gives Holmes the conceptual bridge he needs to describe the intimate texture of Muir's emotional (as well as intellectual) response to natural environments. He takes "relationship as an irreducible human reality," and describes individual development as a process involving "inner images or representations of the objects of the external world," "the subjective experience of those objects, shaped by want, need, expectation, and imagination as well as memory in all its forms." These "objects" are lively, of course. Holmes's elaboration of object relations theory—like psychoanalysis generally—understands "object" to include people: "each person (at each point in his or her life) has a definite array of other persons who serve as 'primary objects,' relationship with whom is crucially important for the sustenance, stability, and aliveness of that person's emotional life."[12]

Primary objects form the nexus of an individual's development in relationship, how one learns to distinguish self from other—and create self with others—in continuing relationship with others, "a means of extending the subject's sense of self and meaningful world into new and larger arenas of action and relationship." Animate and inanimate "objects"—people as well as things—each "carry some of the psychological weight of relationships with loved ones," and inanimate things "may possess a certain psychological life of their own, . . . not as a set of discrete things but as a web of relationships, continuous with the webs of human relationship that constitute family, friends, and society."[13]

Through Holmes's effort, it is not a difficult leap to say that this understanding of individual-in-relationship "can be used to understand the meanings and dynamics of human relationships with the natural world"—Holmes's primary concern documenting the development, not so much of John Muir's *ideas about* nature, but his *relationship with* the natural world in the context of other relationships. His extension of object relations theory allows us to "explore the full range of natural realities—specific animals, plants, landscapes, lived environments, and other natural phenomena up to and including 'Nature' itself, as an encompassing symbol or generalized reality— . . . as primary objects . . . in an individual life."[14]

We can even give up the loaded term "object" to see all parties of relationship as active participants, selves, engaged in responsive connection and development—not just of "selves," but of the very fabric of relationship. From this point of view, a static object would be an anomaly; the standard ground for identity, relationship, and development would be mutually responsive *subjects*, human and otherwise. In other words, the common environmentalist argument that nature is active and responsive rather than passive can be understood, not so much by an intellectual leap of faith or a shift of "paradigm," but perhaps through a more intimately known experience of any relationship. I doubt very much anyone's life or environmental worldview can be transformed merely by having it asserted repeatedly that nature is "active," without some more fundamental understanding of how one learns about and experiences anything or anyone (including oneself) as active quite close to home.

If we understand knowledge, identity, and growth to develop in relationship with lively objects embodying symbolic meanings, it is inescapable that the relationship between oneself and whatever one is reading about is part of the web of relationships under consideration. Holmes understands this, too. He cannot *not* include himself in his biography of Muir. Muir is as much a "primary object" in relation to Holmes as anything else might be, as Holmes's book itself could be for its readers. Holmes certainly creates what we might call plausible deniability—the book is "about" John Muir, and we all know Muir is important: "For scholars and the general public alike, John Muir (1838–1914) has come to stand as one of the patron saints of twentieth-century American environmental activity, both political and recreational." But reflection is built into Holmes's subject as well as his method. Muir understood his own experience to be a model for other people's relationship with the natural world. Holmes writes, "More than any other comparable figure, Muir's influence has been expressed in a series of vivid images of his personal relationships with particular places. This is not by accident; indeed, one of his own primary literary tactics—as well as a recurring rhetorical strategy within the environmental movement—was to offer his life story as the embodiment of a more generalizable model of a certain sort of personal experience of the natural world. Clearly the tactic has worked."[15]

And it has worked on Holmes. While Muir is the occasion for the book, Holmes describes what is at stake in the book in intimate and self-referential terms:

> [A]ll the bright and textured facets of what we call life—meaning, color, feeling, thought, word, story, memory, hope—are made real in the fires of one's own action of body, mind, and heart. This process is perhaps ineffable (in the mystical sense), or fundamental in a way that reaches beneath psychological or disciplinary categories and into those depths of life that can only be understood wordlessly, through living, and sometimes not understood at all. I want most of all to respect and honor these depths, writing not only to analyze [Muir] but to express my own wonder and hope that human life includes such moments of encounter, strength, and creativity. . . . It may well be not through theory or analysis that we come to our deepest understandings but through the felt companionship of another person's life story, as a spur and guide to reflection upon our own.[16]

What brings us Holmes's Muir is empathy, "the felt companionship of another person's life story," and along with it comes Holmes's own reflection and an invitation for us to embark likewise, not just into the facts of Muir's life, but into a personal reckoning of how, where, and through whom we learned to be who we are. He writes, "I offer a story in which we can see Muir—and perhaps ourselves—through new eyes." Not by living the exact ideas and experiences Muir had, but by recognizing the patterning growth of this individual (Muir, and by extension Holmes and his readers) in relationship with people and the natural world, which Holmes can offer us because he recognizes these forces in his own life. What Holmes invites us to do is "not to follow but to walk beside, to converse with, and to learn from" Muir, ultimately to "cultivate our capacities for intimate, mutual companionship with all of the beings that inhabit our worlds."[17]

Though Holmes's theoretical exposition at the end of his book discusses relationships with natural things and environments as part of individual development (by no means uniformly positive, just as not all human relationships are uniformly nurturing), the book strongly suggests that it is in relationships with *people* that the range of possible relationships with nonhuman nature begin to form. They shape what is possible to feel and think about "nature" (and the world as a whole) for better and for worse. The book ends with us walking and talking with Muir (and Holmes), not outdoors by ourselves: "You don't have to be a John Muir to love and care for and dream of and draw strength from the natural world. Muir himself wasn't even a John Muir, for most of his life. You—I—don't need wilderness, solitude, adventure, etc., though that can often help. Rather, you and I can claim and celebrate our circles of human and natural friends and loved ones, and expand those circles."[18]

What we probably do need, somewhere along the way, is a companion—not (just) a mentor to point us in the direction of the outdoors and teach us what lives in the world, but someone to point us in the direction of *mutually responsive relationship*. When we "cultivate our capacities for intimate, mutual companionship with all of the beings that inhabit our worlds," as Holmes says we can, we are cultivating a human capacity. I can't read the word "companionship" without hearing human companionship as its most fundamental reference—even if we come to learn many of its nuances beyond human interaction, with the animals we still grow up with from time to time.

There is a house of mirrors here beyond the scope of this chapter that deserves sustained attention. Companionship in the broadest sense—human, nonhuman, and interspecies companionship—like Bateson's understanding of "mind," and the agency of nature, is more than a human projection onto nonhuman beings. As products of biological evolution, there is no reason why companionship, mind, and agency are not part of a legacy shared with other living (even in some profound way nonliving) things. At the same time, humans are raised primarily among humans; how much or little of the nonhuman world becomes part of a person's immediate, comprehensible, and responsive world of relationships varies widely culture to culture, and from one historical period to another. Obviously what all these relationships mean varies as well. How much did your upbringing prepare you to acknowledge and respond to the voice of a stone or the generous help of a bird *in its language*? We know, however, that many people have had this capacity, a potential virtuosity of being outside the domain of one's species usually presented in cartoon form as "Indians' oneness with nature." What's at stake is nothing so limited as a "paradigm," or an "attitude" to nature, but forms of human being. In the language of white scholarship, I don't think we know nearly enough about what it might mean that nature is structured by "mind" to write off cross-species (or other extrahuman) communication and companionship as "social constructions." I think this was why Bateson was drawn to the resonance of the sacred and the aesthetic by the end of his life—these are profound matters.

Contemporary environmental scholarship and writing suggests many of us are in a rather primitive state of either acknowledging how complex nature is, how far-reaching human effects are on the nonhuman world, or how formidable and long-lasting human imaginative constructions are, perhaps dimly appreciating that some people have known more or known differently. There is a very great deal more. In considering human-environmental interaction, it makes sense to me to attempt to come to terms with the dynamics of relationships where we still have a marginal grip on how these work through actual experience with one another, in addition to attempting cultural openness and responsiveness, discrimination and mirroring. I have no faith at all that more "knowledge" about this or that species, this or that ecosystem, this or that "traditional society," *without reflective development in individual relationships,* will change the orientation of societies that endanger these things, each other, and their own members.

There is no reason it has to be an especially famous or accomplished person (with a full documentary record) that kindles curiosity, self-reckoning, and growth; there is no reason, if one has learned to look and listen with someone—encountering them, oneself, and all the objects of the world—not to look and listen for others (even nonhuman others) in the same way, remote and fragmented as they may be. There is no reason that these relationships should yield only scholarly insight.

Moreover, in an era when global "interconnectedness" and ecologies of all kinds linking individuals to each other in economies, landscapes, and political conflict are invoked regularly, it would seem wise to understand the lived developmental fabric of "connectedness." Therapist Terrence O'Connor has, "upon occasion, interrupted a client's obsessive, self-absorbed soliloquy with, 'Are you aware that the planet is dying?' "[19] He doesn't report why this disruptive intervention seems appropriate to him at the moment it comes up, nor does he explain his own preference for abruptly invoking the largest imaginable grief—significantly his, and not his clients'—in a moment of impatience, empty-handedness, or boredom. He doesn't describe this intervention's immediate effect on his clients in the therapeutic relationship itself, either. What *does* move people, emotionally as well as intellectually, to fully inhabit empathic relationships of all kinds at every scale of interaction (even professionally)? Without attempting to understand that, appeals to empathy or responsibility with anything as large as a planet, or the suffering of its people, seem moot. Likewise, understanding anything as intimate as curiosity and response—the experience of being moved to relationship—demands a look beyond the objects we read and write about, to ourselves. Object relations theory and its extensions into environmental thought is surely not the last word on the matter, but it is a suggestive way to link experience of human relationships with environmental relationships, as Holmes does, self-reflectively, open to the felt companionship of another person's life story.

This is how I approach Ruth Nelson's experience, as a "primary object," the imagined (but still felt) companionship of the fragments of her life story I know, recognized sparks igniting memory of my own, illuminating things not visible before, allowing me to arrange and reconsider what I know, and learn.

Ruth was called back to Colorado and the West and a career as a naturalist. Because Colorado was conventionally beautiful? Something peculiar about the Colorado Rockies? Because she had enjoyed vacations there? Because her informal education outdoors was, as her mother believed, more valuable than formal education? Perhaps. But after a rootless and fragmented childhood (not to mention education), she appears to have been looking for something. However compelling the Colorado landscape may have been, her connections to it were through memorable experiences *with people*. Her parents separated there. The wildflowers around Estes Park adorned her mother's birthday cake. Finishing credits for her degree in Wisconsin, a family she met who clearly befriended her also gave her her first paid work in the summer, again in Colorado, a place she already knew. Her camp experience after college, and her own coming of age in girls' and women's educational institutions, likely formed the idea of establishing a camp of her own for girls. Though that didn't happen, she stayed in the area and brought it into her life through park service work, graduate study and publication, and marriage to a man who had done essentially the same thing (for different reasons) in Laramie. She also collaborated and traveled with women friends often. A lifelong girls' camp on the road, these connections gave her companionship and the illustrations of her guidebooks by Dorothy Leake (a conservationist in the Ozarks, who worked frequently with girl scouts) and Beatrice Willard (a conservationist in Colorado), both of whom had earned PhDs in biology—professionals, and friends. She kept her Skyland Ranch while she lived and traveled with Aven, in Laramie and later in Oklahoma. She returned to Colorado with Aven when he was elderly; it was her most permanent home.

Ruth created company, continuity, work, and home through people, in particular landscapes through plants—her mother's interest—a network that extended over a wider domain as she refined her professional skill, traveled, and wrote, all of it, in her case, in the West. I hear the links of this process at work in Orra Phelps's story (placed in the preceding chapter, "Album"). I hear them as well in butterfly biologist Robert Michael Pyle's story, recorded in *The Thunder Tree* (1993) as he found and explored the High Line Canal near Denver, where he grew up in the midst of suburban landscape disruption and the disruptions of his family's life. As a grown man and a scientist, the seams of work, play, companionship, and experience of place come together in his life in southwestern Washington: "Biographers have noted that Thea, Tom, and Dory [Pyle's third wife and her children] helped give Bob a stable, orderly family structure for the first time in his adult life," a structure that embraces and allows his growth as a writer, a naturalist, an activist (and a person), through which he inhabits and responds to the landscape he so vividly describes in *Wintergreen* (1986), his "breakthrough literary work" published soon after his marriage to Thea Hellyer, a woman he'd known since college, in Washington.[20]

Why do I hear a persistent echo of the importance of relationships in specific places in the absence of a long scholarly "literature" on the subject of naturalists' (or anyone else's) affect, family life, memory, and choice of residence and work? Because there is a loosely suggestive similarity between these people's histories of disconnections and reconnections, including their relationship with particular landscapes, and my own. Because of my differences from any naturalist, I am curious about what sorts of mirroring and empathy are going on here. This curiosity has a history, bound up with how I learned empathy, awareness of the nonhuman world, and the most basic forms of reading, interpretation, and response, in relationship with other people. I offer it here, not because the details are themselves important, but because each one suggests the possibility of empathy and recognition in the presence of others, including Ruth Nelson. They created the possibility of listening for her in very particular ways. Each piece is an image, a "primary object," a question and a response, in a conversation over time acted out between people in long chords of learning. Some of this learning is "about nature," but really the whole fabric is about human being in relationship.

*A note passed between two girls in a Normal School classroom in East Randolph, New York, in 1917:*

> Did Helen and Walter drive the car the A.M.? What a relief it will be to get out of this stuffy old school room. My head is nearly cracking open. Too much sound sleep last night I think.
>
> *[the reply:]* Sound sleep! Eh! What! Ha. No they came with a four legged

"Ford." Yes, this is stuffy. Every time I open a window it gets closed. Audra likes Fredonia.

These are the voices of Hazel Ellis and a friend, teachers in training. Helen was Hazel's younger sister. Audra was at the teacher's college at Fredonia State, a little older and a little more ambitious than these farm daughters. Automobiles were still a novelty, and the girls were still very much girls. Hazel was sixteen. Tucked into a notebook crammed with perfect penmanship recording the day's wisdom on teaching spelling and grammar, history and geography, physical education and science, along with a small sprig of a wild rose and homework assignments, this little surreptitious communiqué in a stuffy classroom on a long afternoon almost a hundred years ago startles me.

The only reason it could is because Hazel was my grandmother; I discovered the note in her school notebooks after her death. A recognition and a distance. She is there in the handwriting, which is youthful but essentially hers; she's there in the classroom which would be her element for the next fifty years; and I recognize Helen and Audra and the Ford, four-legged and otherwise. And the wild rose. But she's bored (in a classroom!), and was up carousing somehow the night before—news to me, about the intensely quiet, shy woman I knew and loved. At sixteen. An age when my own carousing brought the wrath of curfews and groundings down on my head, usually in connection with a medley of suspicious boyfriends. My grandmother watched mostly without judgment, holding me in a warm distance from the immediate fracas—maybe, I think for the first time, in empathy. She had the luxury of boarding in town to go to school, even for high school. I wouldn't have that until I left the suburbs of Buffalo to go to college.

Hazel was the first child of a very young woman. Ida Pearl Seaton married William Ellis when she was sixteen and he was in his thirties, at the turn of the last century. They were tenant farmers, and moved west across New York State from Oswego, through the districts burned over by religious revivals and assiduous farming. They bought land and built a house in Napoli sometime in my grandmother's childhood—not right away, another daughter and two sons were born in other houses around Cattaraugus County—and raised some things to sell, many things to eat. The children slept two to a room, Hazel and Helen in one and Cleo and Lynn in the other, off opposite corners of an odd upstairs sitting room, where there was a stove. My grandmother's lack of privacy guaranteed my mother her own room, so the story went, though she was an only child. In winter, the girls' feet were so cold they rubbed them with snow on the windowsill to soothe their chilblains. In summer they were barefoot; expensive shoes were reserved for the school term. They tatted their hearts out, long lace webs of tiny knots in thread, and covered their beds, their chairs, and their dressers with acres of crochet and difficult beautifully pieced patchwork. Ida's husband, Will, died when my mother was a little girl; she remembers him as affectionate and funny, a contrast to Ida's humorless determination, which I knew. We know nothing about their marriage besides that it took place between poor people, one a girl, probably with few options for leaving home or supporting herself. She was slimly literate and kept her daybook diaries in her spoken idiom—fragmented, distinctive. "Lightning come in the house today and blew out all the fuse."

There are no stories revealing the emotional geography of this family—a few pranks involving outhouses, memories of the old world without running water, electricity, and central heat (in which my great-grandmother continued to live for the most part)—but this fact, coupled with how all these children's lives ended, speaks volumes about failure of some indeterminate kind. The boys' lives were pretty much erased by the time I was old enough to understand these people as my family, even though Cleo at least was alive until I was three (no one ever talked about him and I don't remember meeting him—I discovered to my surprise in my great-grandmother's diary that he died in 1966). I do know that they all married and they all found work. Helen worked for wages at Borden's creamery in Randolph. (So did her mother, on and off, after she was widowed.) Hazel and Helen saw each other often. I remember Helen as an invalid, and was surprised she was alive (being the little shit I was, once told her so), lying on a hospital bed in a floridly papered apartment with impossibly high gloomy ceilings, addicted to prescription drugs. (The doctor who gave her this stuff was the family doctor for years; she delivered me into the world, and was another grandmother, stepmother to my father and his sisters: Dr. Ruth Knobloch, Doctor Mom, Gramma Doc, another story of fissures and connections.) The boys' lives, like Helen's, did not end well. One drank himself to death and the other shot himself in the head. The sturdy young men holding the halters of their strapping workhorses in 1920 only exist in a photograph. Ida was the one who stayed on the farm; the rest fled, and for the most part unraveled.

Hazel looked out for her mother and sister, and for herself. She propelled herself into teaching as soon as she possibly could. She kept her job after she married in 1922, which was somewhat unusual for women teachers of her generation, and perhaps because my mother was not born until 1940 (Hazel was thirty-nine), it was unthinkable by then that she would lose or leave her job for motherhood. She hired a younger woman to live with them and care for my mother (Edith Nichols, "Edo" to both my mother and me). Though teaching was a necessity for a seventeen-year-old girl leaving home with some measure of independence, this was something Hazel loved and did well. She was patient and orderly, and gave her students—many of them from farm families like her own—an environment for learning that was demanding but also engaging and responsive. My mother's birth was longed for and elusive, but it could as easily have been planned that way; it's possible, I think, that Hazel had no intention of repeating her own mother's early and maybe unhappy or disorderly motherhood. She never talked about it, but my grandmother's life has all the hallmarks of an extraordinary will to self-sufficiency and self-preservation. And she married for love, not security. Her husband, Harold Fargo, was a farm laborer when she met him during World War I, unfit for conscription because of his poor eyesight and more valuable on the farm. He later made a living fixing things and working in McNally's hardware store in Randolph, and later still kept the school swept and its plumbing and heating intact. He never finished high school. Tinker. It is very likely that her income exceeded his all the years they were married. Harold made her laugh, and made her "angry," calling her Huzzel, which she claimed to hate (though when emphysema took him from her nine years before she died, I'm sure there was not a name in the world she would rather have heard).[21]

Something allowed her to imagine a full life apart from the farm and her family—away from whatever failures and shortcomings I will never know—that her siblings never found. What? She stretched and genuflected in the genre known as "How I Spent My Summer Vacation" in the fall of 1917:

> After school closed and commencement week was over, I was ready to make plans for my summer vacation, as I had made none before school closed, but as it did not turn out as I planned it should, I will endeavor to describe the way I really spent the summer.
>
> As my school work was especially hard last year, my first thought after school closed was to take a rest. For two or three weeks all that I did was help mamma with the house work and worked a very little in the vegetable and flower garden. The rest of my time I spent in taking long walks

through the fields and woods, going for automobile rides or sitting on the veranda on the hot summer afternoons, crocheting or tatting which are my favorite pastimes.

After spending two or three weeks in this manner, I felt much rested and ready to do my part again. As it was now time to harvest the hay crop and help was very scarce, I decided to help with the haying, so for two weeks I took the place of a man in the hay field. Although this was rather hard work for a high school girl to attempt, I stood it very well and enjoyed it very much as I like to be out of doors where one can see and enjoy the beauty and wonder of nature.

The next day after haying was finished a lady, who lives on a farm not far from my home, wanted me to come and help with the house work, so I went. There I stayed four weeks and it was then time for school to commence again.

Although very little of my vacation was spent in pleasure, I must say that I enjoyed it the most of any summer vacation I have ever had yet.

She got a "B" on the assignment. As tidy a set of lies, evasions, and sublimations as you could imagine, her lines of flight are still clear. Most obvious is the fact that the lanky stylized school prose is a far cry from her mother's daybook notes. Her school work had been hard because she had graduated a year early. That plan was obvious; she wasn't allowed many others of her own. "Rest" was gardening and helping her mother, who, it would appear in all memories of her, was irascible, driven, and demanding—maybe her mother thought she was lazy. The "veranda" was a long scrappy porch. Help in the fields was scarce because there was a war on; Hazel most certainly did not "decide" to help with the haying, or to work for a month in a neighbor's house. She was not such a clever liar that she'd transform the work of the summer into unmitigated joy, but admitting that little of it was spent in pleasure, she still enjoyed it more than any other summer—an odd point to make in a formulaic essay. She could as easily have left out the whole question of disappointment, failed plans, and the difference between "vacation" and summer experiences that were good for her.

Why did she end the essay the way she did? I think because repeatedly she was out and away on long walks and rides, away from the house, very likely with other people. Moreover what she learned in school (with female mentors she talked about all her life and obviously identified with) would have promised something rewarding outside, with friends; her school curriculum was saturated with "nature study," and her social life full of other young people, many of them girls her own age studying the same things. The "beauty and wonder of nature" as well as pride in her work in the "place of a man" (alongside other people) made the haying *feel good*, and in the end she left her family entirely for a month (to live with still other people). It was a good enough summer, and not a "vacation."

I learned "the beauty and wonder" of nature and the pleasure of being outside with her long before I ever saw this little essay, in a similar (not identical) way. It did not occur to me until I ran into Ruth Nelson that Hazel might have wanted—needed—to leave her house as a girl. I suspect pretty strongly now it occurred to Hazel that I did, too.

My parents separated when I was little, in a grinding wreckage of young people's misjudgments and cruelties. My father's flight was towards graduate study and another woman. I remember sound and fury, and long years of my mother's heartbreak. A break of my own reverberated everywhere, in my head, my heart, and my body. Ever since, I have had an intense longing and appreciation for anything that is not recklessly sundered by myself or someone else, including places inhabited by nonhuman beings—this was my grandmother's contribution. Not a full-blown "environmentalism," but the potential for it, because "the environment" was significantly charged with my memories of her.

Home, meanwhile, in every way that seems meaningful to me now, was really with my mother's parents, I think maybe for my mother too. It is where I learned what "home" might mean, there or anywhere: a place, a set of relationships, things to do and be curious about. My mother was ambitious to get out of elementary school classrooms with a graduate degree in psychology, to challenge herself more, and support us better. We had lived with her parents in Randolph when I was born, and for a while afterwards when my father was in the army. After the separation, my mother and I went to Randolph often; when she was in graduate school, of course I would live there.

I spent a lot of time in Randolph outside, winter and summer. My grandmother took me on long walks through the fields and woods, and long drives in the automobile, introducing me to the companionate experience of landscapes that I strongly suspect she learned to enjoy as a girl. I learned what lived "out there" at the same time that I learned, just as precisely and through the same five senses, how close she was right here, what "close" meant. Lessons in bird calls and animal prints, the shapes of leaves and the texture of bark, were lessons in listening to her voice, watching her face, touching her hands. Through her, there were winter cardinals and summer buntings. Marvelous blue speckled eggs in tidy nests, and fragile birds fledging from her birdhouse. The seasons of flowers wild and tame. What lived outside was a map of my relationship with her before it was "the outside." Had she been a farmer, I would have learned that too; among other things, I learned some of the unconscious fabric of what teaching is.

My grandfather taught me more about other kinds of interactions, both close at hand and outside the family. He pulled me in a red wagon from one yard to the next so I could get petals for a perfume "factory"—roses, peonies, gladiolas, clematis, lilacs, iris, more I can't remember, all perennials, and all planted over a period of about forty years by my grandmother and Mrs. Bowen next door, melding their yards together in a froth of plant life, trellis gateways, and permanent friendship. Our vegetable garden was on Mrs. Bowen's property, a singularly irrelevant fact. I helped my grandfather weed there and pick vegetables. On early morning tours, with the smell of crushed creeping Charlie in the air, dew soaking my sneakers, my grandfather bantered with neighbors. The old men called out to each other across the yards, and teased each other about arthritis and their stupidities; they talked weather, and flayed politicians. My grandfather hung a swing in the old apple tree, which was useful picking apples by bouncing in it, bringing the sour little things down on my head in waves. My grandmother tried it out (in her eighties) and found herself on the ground laughing with the apples when the ropes broke. He gave me a bucket to hang on a tap for maple sap in the spring, and took me with him to burn trash in the fifty-gallon barrel at the end of the long yard, "out back," with the compost heap (also on Mrs. Bowen's property).

I spent many hours alone out back, but this was a deeply known universe even when I discovered something new, like how it felt to fall fifteen feet out of a tree flat on my back, hearing my grandmother call a backyard or two away, being unable to breathe, thinking very clearly, "how strange!" and, after I staggered into the kitchen, learned this was called "having the wind knocked out of you." Widely applicable. A blow, a long breathless voiceless pause, and the likelihood (not certainty) that all would be well, and the day would go on. Everything was not perfect—this place was not a closed sanctuary against what I feared most. But this world was where things, including myself, did not come permanently, violently apart, and I had the luxury of minute, tantalizing continuities. It was my grandparents' universe. I knew all the neighbors and they knew me. The far edge of the world was Lorana and Harry Rasey's house, and across the span of yards were the many years of teaching Lorana and my grandmother had done together, nights of bridge and pinochle, afternoons picnicking and finding wildflowers, watching children play, serving creamed new potatoes and peas just out of her garden. Both Fargos and Raseys are now buried side by side in the Randolph cemetery. The miles of farmhouses, all named for families who had lived there, Virginia creeper, corn fields so humid they must have been breathing, roadsides and pastures full of Queen Anne's lace and blue chicory, the graves in the cemetery we visited each Memorial Day—all this was real because these people were real and continuous to me, even in death, and I had a home with them. The sound of a blue jay calling over the yards from Lorana's tall spruce trees, arcing over the known world, is still the sound of home. In the voice of a mountain jay, I hear it in Laramie.

I knew much of this story long before I met Ruth Nelson, but only as the murky background of an emotional life that had little to do with official things like school and work. In fact, just the opposite: outdoors was "away," away from parents, teachers, conscious life in general. This is a long habit, and not just mine. Thinking about Ruth gave me a chance to revisit what I remembered about learning to distinguish among flowers and plants, with whom and under what circumstances. But as I read over and over through the Nelsons' material, unable to let them go, unable to explain why Ruth Nelson had become more interesting to me than her husband (how?—there is so little left of her), the faint trace of her family's disruption that Ruth left in her interviews finally fell like lightning on the fact that she worked in Rocky Mountain National Park. I had been there. I live not two hours from there now; I had forgotten the entire event for the better part of twenty-five years.

When I was in high school, bored in my tenth-grade chemistry class one afternoon (taught by an intolerably difficult man to whom I promised I would never be a scientist, though I took college chemistry from him two years later, a long purgatory), I lingered over a flier he had passed around advertising a summer field study program through the University of Iowa—spend a few weeks backpacking and studying in Rocky Mountain National Park in Colorado, and earn one credit of college biology. I really wanted to go. By this point, my mother had been remarried for some time and my two brothers had been born, our home life was materially comfortable but emotionally very tense (to me, most of the time unbearable without a serious deadening of every response I might have to it), I both loved and hated school, and was deeply entrenched in romantic habits that would plague me for a long time afterwards. But I had also been raised as a keen if intermittent amateur naturalist. This summer trip would get me outdoors, far from home, with the justifiable value of being educational. It was the first time an inchoate comfort, curiosity, and pleasure anywhere, in this case outdoors, might have some connection to things more obviously "useful" (the next was when I married and moved to Montana, and the last was when I moved to Laramie). My parents bought all the gear and paid the tuition.

I am embarrassed still by the barely conscious, reckless girl who took that trip (and the woman who did not remember it), and she was not inclined to see how her pleasure in the field with partners and pals might add up to work like that for real—she wasn't brave enough to mark out her own education, and went on to study in a general way what her father studied. She was too socially awkward to have any grace among strangers, and too scattered to know she might be learning something. Still, plotting out all the living things she could find in a tiny rectangle somewhere in the backcountry, utterly satisfied with a dwelling in a backpack, she started to trace the line of a life that would be fuller, and different, later. I told my grandmother proudly I'd been studying what lived around a "meandering stream," and started to tell her what this technical term meant. "I know what a meandering stream is," she said. If I were to plot that rectangle now, I would include the girl.

Looking for Ruth Nelson led me back to her. Some of the reasons I moved to Wyoming I sketch out in the preface. Ultimately all of them converge around a constellation of work, family, memory, companionship, pleasure, and a particular landscape, where the peaks of Rocky Mountain National Park are visible, and my first western home is a day's drive away. Still leaving and finding home, drawn close to a landscape I had been prepared for and drawn to repeatedly by how I learned who I was at home, Wyoming made intuitive sense.

Life in Laramie wasn't comprehensible or satisfying right away; teaching flourished but writing was very difficult. I was busy but also gravely disoriented. A young gay man, Matthew Shepard, was murdered a year after I arrived. Guns, alcohol, a very rough edge. Among other things, I sought out Quakers and a therapist. Wind knocked out, I was listening for a call. Through work I opened questions about the place I was living that I couldn't look into any other way; a literally familiar curiosity led me to a local botanist, and then his wife and their lives. My education long since closed over a career as a naturalist. But the work I could do, reading for professional ends, rattled what became an almost audible resonance between my desire to live here and what I wanted my life and my work to look like, my curiosity about this place and botany and these ordinary people, and the questions I asked of them—what was it like to work outside exactly here, seeing and listing plants the way they did? What about Wyoming and Colorado made this work possible and desirable? What was it like to work and play with friends and a partner, all in the same place, and *with* that place? These people accomplished something immensely valuable and intellectually compelling. Though it was the very last thing I asked, trying to listen to and learn from Ruth Nelson especially presented an obvious question: are these things possible for me here?

I think it's plausible that botany in Colorado and the West meant something particular to Ruth because of her family's experience there—a sundering, and a reconnection. I think it's further possible that in a life marked by disruption in family, residence, and education, she would seek continuities somehow. Her circle of women partners and friends, and her idea of starting a girls' camp, were surely the result of her experience in girls' and women's schools. A form of intimacy with the western landscape allowed her to respond to her mother's interest in plants and at the same time transcend her parents' difficulty. Her late marriage to a man who would work with her both allowed her to master the scientific work of botany, and may have clarified for her that her interests were really elsewhere, in description, appreciation, and possibly a simple but full well-being. None of this salvaging and forging of relationships would undermine her work as a botanist; on the contrary, her choices appear to have brought threads of her experience and memory together in a life in which each part spoke to other parts, creating herself as well as her work in ongoing relationships with people in a particular place.

I have wished that she had been, in print, a more public champion of the Front Range open spaces she saw covered with new suburban houses; she offered recommendations for landscaping without noting that these sprawling developments would endanger the very "wildings" she spent her life with. But that's my concern, not hers. She left us field guides, and the experience of her students teaching classes from Rocky Mountain National Park after Aven's death. The *form* of her work says, in a much more muted way, where a primary relationship with natural landscapes and living things might come from. More than that, where work and home might come from together, in a very ordinary life. I am not referring to the particulars of family malaise, though that was exactly what I heard first, because such stories map deep contours and fault lines people negotiate with or without transcendence all their lives, and I can't help being moved by them. Laying my experience side by side with Ruth's, my grandmother's, Orra Phelps's, and Robert Michael Pyle's (and Steven Holmes's John Muir), it is the role of relationships with people that strikes me in each. That "nature" has also been an object is, I think, if not secondary, only one reason to venture into this territory.

Actively forging connections between intellectual work and the rest of one's life—memory, present experience, place, emotional life, as if these things might be lived and develop together rather than separately, in close dialogue with one another—poses risks, especially professionally. It is a commonplace that contemporary heirs of Western civilization compartmentalize all things, their engagement with others and the world at large, and even themselves. Revisiting how the sundered parts of experience might be reconnected is not an effort of losing the power of distinguishing between one thing and another, but rather can be an effort to make different kinds of distinctions visible, to look for parts of one's life and others' lives we rarely allow ourselves to talk about.

I know I have read Ruth Nelson through memories and desires of my own, projecting and introjecting, but I am conscious of our differences, and the encounter allowed me to learn something *from* her, not just about her. It was her book I took outside to make sense of the floral world I had moved into, and her habits in the field notes I became captivated by. I understood how important botany ("work") was in the life of her marriage when I could read plant names sufficiently in the context of the Nelsons' travel to see that they had recorded vacations—even food and bouquets—in an otherwise very scientific-looking record. Her approach to collecting was different from her husband's: he listed scientifically, and left to herself (significantly, after he died), she also described aesthetically. This difference is what would allow her to complete both lovingly *and* professionally his life's work in the course of her own, in a scientifically informative but popular guide to Rocky Mountain plants, the "dual purpose manual" he never wrote.

All these facts may or may not contribute anything to the history of botany, but that has not been the most compelling context of Ruth Nelson's life for me. It was a relationship between us—attenuated, mediated—that revealed a whole concatenation of relationships that made botany meaningful to her (and then to me in new ways), and suggested relationship itself—human and empathetic, responsive and nurturing—as a cornerstone of many kinds of continuity and knowledge, including an understanding of the natural world.

Left to right: Neva, Helen, and Celia Alice Nelson at their home in Laramie. Courtesy American Heritage Center, University of Wyoming. Aven and Ruth Nelson Papers.

Ruth and Aven Nelson and Medicine Bow Peak, 1931. The Nelsons composed this print block the year they married, possibly for their wedding announcements. Courtesy American Heritage Center, University of Wyoming. Aven and Ruth Nelson Papers.

Willard Ashton's Horseshoe Inn, with pond. Courtesy Estes Park Historical Museum.

The house in Napoli, New York, built by my great-grandparents. Author's photograph.

Ida Pearl Ellis (my great-grandmother) and a friend. Author's photograph.

Hazel Ellis (my grandmother) on the occasion of her high school graduation, 1917. Author's photograph.

Linaria vulgaris

Bog rein orchid

Sickletop lousewort

Meadow rue, female

Ceanothus

Timothy

Epilobium angustifolium

Iliamna rivularis

Flax

Geranium

Lousewort

Elephant head

Silky phacelia

Lupine

Nettle leaf horsemint

Penstemons, blue, purple, white, striped

Phlox

Rydbergia grandiflora

Mimulus?

Castilleja, brick orange

Yampah

Moss campion

Spring beauty

Showy gentian

Moss

Fatigue

Heartleaf arnica

Hotrock penstemon

Galium boreale

Habenaria dilatata

Smelowskia calycina

Dryas octopetala

Perideridia gairdneri

Silene pratensis

Wyethia helianthoides

Potentilla nivea

Lonicera involucrata

Lonicera utahensis

Anemone multifida

Pedicularis bracteosa

Aquilegia caerulea

Senecio serra

Castilleja sulphurea

Apocynum androsaemifolium

Collomia linearis

Pedicularis groenlandica

Rudbeckia occidentalis

Townsendia alpigena

Prunella vulgaris

Hackelia floribunda

Valeriana occidentalis

Penstemon montanus

Penstemon whippleanus

There is no sense in searching
for the secret of what anyone
may have known.
—Jacques Derrida,
*Archive Fever*

# Collecting

*I notice his age first, because he is thirty-six, and so am I. His clothes are formal by my standards—vest, trousers, jacket. Sensible boots laced up above his ankles. He looks midwestern, sturdy and blond, scholarly now, and slight. He seems comfortable here. I won't disturb him. He is curved over plant specimens under a lamp at a dark desk. Nothing in his office matters outside his gentle attention to his work. He looks like he knows what he is doing, glass in hand, poring over the petiole of a leaf. He will not look up though I watch a long time near him. He can't talk to me of course. There is only his posture to read, the hour, the crossing and recrossing of knees in coarse wool, the slow circulation of tools in his hand—glass, pen, paper, book, spectacles. I am here to see him work, to see his work, loose mounds of specimens on heavy papers, his summer's collection dry and flat, turned now this winter to a long season of study. No, he does not know what he is doing; the pen goes down. A small crisis of a private hour. He meets it without disappointment, impassive and quiet. He reopens Gray's* Manual. *This isn't quite his work yet. All his references lie heavy as Bibles around the dim edges of his lamplight. He is sure only about the fact of this plant on his own page, and the day he plucked it whole from the world, but here taped to its paper it is mute. He needs its name. He is not Adam. Both of us are at a loss. He has to read Gray; I have to touch those plants, the cool gritty soles of those brown boots.*

*I haunt Laramie, a breath I barely recognize as myself, free, lost. He goes about his business in the Rocky Mountain Herbarium, piling specimens into cabinets and new species into botanical science. I can't leave him alone. I follow him to the vacant lots, through the scrappy pasture now manicured in the middle of campus, up Medicine Bow Peak where the asters bloom improbably late at eleven thousand feet on my birthday in the fall. He doesn't notice them when he races down the mountain to his dying wife. Though this is a mistake, I try to make him speak from his letters, his articles, his sermons, his lecture notes. Tell me where I am. Tell me how to live here. Brittle history appears on my papers—about Wyoming, about botany—as I sit curved over my own desk. He hasn't said a word.*

*But he is speaking all the time, through the soles of my own boots, my eyes, my hands full of soil and leaves, the thin air in my lungs in an alpine swale near Brooklyn Lake. What no one can tell me directly I learn by physical poetry in space and time, his loose rhyme cradling the entire Laramie Basin and my life in it. He lived and worked here. His heart and body moved through these campus buildings, and across the basin, where I can also go. A slow alchemy makes me visible. I am lucky; he has left a box of sedges behind in the herbarium, unidentified, for a hundred years. The slender stems are still green when I see them. Touching them my hand becomes real.*

AVEN NELSON AND Celia Alice Calhoun Nelson had been married for forty-five years when Alice died in 1929. At the top of Medicine Bow Peak, the highest rise of the Snowy Range Mountains west of Laramie, Aven received a call at the lookout tower that his wife had fallen seriously ill. He had just led a group of visiting botanists up the rocky scree of the ridge that early August day, from which they would be able to see across the Laramie Basin, the little lakes nestled in its floor, and the bright ranges to the south that include Longs Peak in Colorado and Rocky Mountain National Park. Although Alice enjoyed summer field trips and accompanied Aven and his colleagues often, she had stayed home that day. Aven's guests had come for a few days' trip organized by the Botanical Society of America, which Aven arranged to begin at the university's science camp a few miles from Medicine Bow Peak. It was one of his favorite places. Aven had taught there in the summers for several years since it opened in 1923. Geologist S. H. Knight and his students cut the logs for the camp's rustic buildings, and built the lodge and cabins themselves. Many of Aven's students remembered science camp with enthusiasm; Aven loved teaching and collecting in the range and went there often with Alice, whether for work or family picnics.

Aven's teaching appointment at the newly opened University of Wyoming brought them to Laramie in 1887, two years after their marriage. He was an accidental botanist. Aven found himself responsible for identifying some plants left behind by the university's horticulturalist, in addition to multifarious other duties. He began his own collecting in 1894 and embraced this new avenue of inquiry energetically; it formed his professional and personal identity for the rest of his life. Throughout his long effort to create himself as a botanist, Alice was his close and understanding mate in life and in field work. She and their daughters, Helen and Neva, had accompanied his exploration of Yellowstone National Park in 1899, an important trip in his development as a botanist, and a memorable summer for the family.

*There is a crowd of acquaintances I recognize in* Wyoming. Erodium cicutarium, *red-stem filaree, shows up out of nowhere in a friend's rock garden;* Convolvulus arvensis, *field bindweed, has been happily smothering grass and now climbs goosefoots and tumbleweeds* (Kochia scoparia *and* Salsola kali) *in my yard. Looking at the long lists of plants, though, that Aven Nelson collected from 1894 on—about sixteen thousand of them, almost entirely Latin binomials—I am awash in a vast language. So was Nelson when he had to get serious about field work. He came home to Wyoming in 1892 with an MA from Harvard and absolutely no experience collecting plants, the very activity that made his publications possible. Nowhere did his Harvard professors speak about plant collection, identification, and preservation. Still, like Nelson, I half know what I'm looking at, especially in his early, local collections. His first season was practice of the most endearing kind. Scouring the Laramie hills and plains, from the Snowy Range west of town, to Telephone Canyon in the Laramie Mountains to the east, as far north as Wheatland, he also tried his hand at plants growing on campus, in Colonel Downey's vacant lot, at the corner of 8th and Kearney Streets, and in the soil of the Laramie streets themselves. He gave as much attention to* Taraxacum officinale *and* Populus monilifera*—the dandelions and cottonwoods we still see every day —as he did to anything else.*

*The floral world is as vivid as the social world.*

IN 1929, Alice's presence was surely missed when the hikes began the first of August; when the call came at the lookout tower, Aven came down the mountain to face a life full of accomplishment, without a companion. In a letter to his friends he said, "For more than forty years, on my return home, from even an hour's absence, if I did not see her, my first question was, 'Where's Mother?' . . . There is no one with whom the incidents of the day may be shared with perfect understanding. Significant or trivial, they were part of the little intimacies of our life. Even when nothing was said, quiet evenings, each busy with familiar tasks, there was a sense of companionship, which exists now only as pleasant memories." He wrote, "Now I have to learn a new way of living."

*There are thousands of pages to read over, but he never wrote down the simplest thing—what was it like to work outdoors? He "loved nature." What does that mean?*

*In search of Aven Nelson's heart in the field, or at least his physical presence in it, I hold his working copy of his own 1909* New Manual of Botany of the Central Rocky Mountains (Vascular Plants). *With the Cruciferae he wrote in* radish, turnip, cauliflower, *linking a botanical family name with plants and especially foods students would know; Rosaceae—*apples, pears, plums, strawberries, blackberries; *Ericaceae—*coffee. *He taught from this book. He dropped pieces of things he held—a leaf, a petal, a bract—into its pages. His pens leaked on it and it sat in water. Its sleek inky cover was rubbed gently fox red in his pocket.*

*His field notes are in every way more stiff. Smaller, easier to carry, each collection book holds the record of a season or several years of collecting. Many are ledger books, as if to count up his floral riches, specimen by specimen, in a serial numbered list. These are the books of the taxonomist at work.*

*The broad brush of the first identification—*Aster, Castilleja, *or nothing at all but the number, always a number—is followed later with a definite name:* Castilleja sessiliflora, *not just any paintbrush, but the Great Plains paintbrush, a loosely built mauve and white cousin of more snappy red paintbrushes. Nelson found it 19 May 1931 near Wagon Mound, New Mexico, on the first day of a two-week collecting trip. He also found* Townsendia exscapa *on gravelly sandy ridges in the same area. The light-pink stemless daisy was blooming low to the ground when he arrived; had he come later that summer, he would have missed it. The names he listed are formal, carefully completed.*

*He rarely tipped his hand with a common name. One daisylike flower he listed only as a "strange comp.[osite]" and identified it later as* Berlandiera lyrata, *which looks like a yellow daisy missing every other petal—he had no idea what it was until he looked it up. But his descriptive remarks are few, just brief notes about location, and a date, very rarely color (he was partially colorblind). Long lists of names are followed only by "do,"* ditto, *something found on the same moist riverbank or roadside as everything else since he wrote the information first. The numbers reflect the order in which Nelson identified and catalogued each specimen in the herbarium in the months following a busy summer in the field. You can't see, feel, or smell the plants on this list unless you know them beyond their Latin costumes.*

*With these diminutive, orderly books in hand, you get a sense that a full floral world thrives busily just beyond the page, hovering in colorful but veiled confusion. Each living plant comes eerily into focus listed in the ledger by Nelson with a Latin signature. Nelson carries his registry through thousands of miles of mountains, deserts, riverbottoms, grasslands, and woods, asking that the crowds he meets reveal to him one by one who they are in the language of his profession. He gathers them, too, roots, stems, leaves, flowers, and fruit, and carries them home flattened neatly in the press.*

AVEN COLLECTED repeatedly in and around Laramie decade after decade, sometimes for whole seasons, as in the mid-1890s, sometimes for a few weeks or only days in the summer. The alpine lakes, meadows, and talus slopes of the Snowy Range were Aven's regular stomping grounds. The Centennial Valley, just below the range, the Laramie plains and hills, Sybille Canyon and the Platte River to the north were likewise frequent summer collecting places. He knew his home watershed very well, a botanist's paradise abundant in wild lives. His familiarity and the intensity of his early collecting (almost three thousand specimens from this area alone in three years) never exhausted his interest in well-traveled places close to home. From 1924 to 1938, when Aven was part of the regular faculty of the science camp, botany field trips were invitations to add numbers to his list. He continued to return to the range after his summer school duties were past, picking up odds and ends as late as 1944, his last year in the field in Wyoming. Even in the 1920s, he was unlikely to find anything unusual in the range—these were not exploratory collections—but he dutifully noted *Salix*es (willows), *Castilleja*s, and *Penstemon*s that he must have seen hundreds of times and had collected before. Aven listed collections as a personal habit. These lists were his immediate record of studying and taking pleasure in the seasons of growth, flower, and seed in the landscapes he knew best.

*At the top of Rogers Canyon Road, near home, with the Leguminosae/pea family. Flowers papilionaceous. I know the French word "papillon." No pods yet. I know the English word "banner," too: this banner is purple at the base, colored as if by fine inked lines spreading and tapering on white ground. Small wing petals, white. The keel is white at the broadest curve, deep purple at the tip that points up into the banner. Banner ½". I refer over and over to a simplified diagram of flower parts, and a glossary, back to this little pea-flower. Involucre is clear/greenish with maroon hairs, cuplike with five delicate points, very small bracts at the base which are green, with hairs. Stem pubescent too. It has six flower heads arranged in a raceme. The leaves are compound pinnate. Stem 1 ½", leaves 1", opposite, with stipules. Odd-pinnate. Either* Astralagus *or* Oxytropis, *milk vetch or loco. I skipped a single species in the key—sweet vetch—leaves too big in relation to flowers, looking at the illustration. I also skipped a category that included alfalfa and sweet clover—this plant is pinnately five- or more foliate, and not a shrub. I can't get the species without the pod evidently. I'd say the tip of the keel is acute, not blunt. So it's an* Oxytropis *and not an* Astralagus. *If, by "tip," one means an end-point, away from the base of the flower. I realize I doubt what "tip" means. So, say it's an* Oxytropis. *"Stemless": it has basal leaves and a flowering stalk. Yes. It's not sticky. Leaves are not whorled. Racemes six- to many-flowered. Calyx with appressed hairs.* Oxytropis lambertii? *Depending on pod I guess. Lambert loco. But the color is wrong. It's "ochroleucus," as the key says, but it's purple/blue too. How about* Oxytropis campestris *var.* cusickii *based on the size of the flower (10 mm), and length of the "scape" (less than 15 cm, no hairy stipules)? I have no idea.*

*Back to the* Astralagus/Oxytropis *key. This one now—an* Astralagus? *Tip of keel is round. Leaves look simple, not pinnate. Not tufted. Maybe* Astralagus ceramicus *var.* filifolius? *Because of the leaves. They aren't compound and it's small.*

*In an hour with these little plants, I at least know the difference between an* Astralagus *and an* Oxytropis. *Beyond that I don't know what these things are; someone else could tell me easily. All I have is this book. They're common—they're everywhere.*

WHEN ALICE DIED, Aven took a trip. He planned to spend a couple months in the Southwest, looking for a plant he had named in 1924 for his wife, *Calhounia nelsonae*, though he had never seen the plant where it grew. He had named it for a specimen sent to him by a colleague. Lonely and at loose ends, he wanted to find it himself.

*I've been in Yellowstone and Grand Teton National Parks for a week. The car is home, and I have enough clothes to keep reasonably clean, a camp stove, a box of canned and dry food, a growing library of plant books. Yellowstone, as the great national outback, is a profoundly social place. Roads go where rivers go, as they did in 1899, and everyone is on them still. I am a hazard to traffic moving at sixty miles per hour (the speed limit is forty-five)—I turn off where it's mostly safe, and gnaw at delays for cruising landships. But parked and settled, I can just be here. Unlike Nelson, I'm not doing anything remarkable; no one sees me. Couples prowl the parking lots and rest areas where water and other diversions lie. All ready to hike up Bunsen Peak and leave a new note in the metal box at the summit, I end up sitting out a storm in the parking lot. Mark and I walked up Bunsen Peak last year. Fallacy or not, the objective correlative holds sway here; many of the burned trees we saw standing then lie like huge jumbled matchsticks now, and those left upright crack and howl in the wind. I found an* Erigeron, *a small flower like a daisy, but it was missing all its "petals"—it had no ray flowers. He loves me, he loves me not. I am still not sure what it was. I also saw the bright show of silky phacelia, a luminous purple fringe I'd never seen, that no photograph I have does justice to.*

*I took the Dogshead Trail into the woods through a four-foot carpet of new pines, walking alone worrying about bears, to see what's become of the 1988 burn. The standing trees are gray, leaning on each other, groaning and creaking. They sound like they might crash down any minute. If you run a branch of a little pine through your hand, a loose handshake, the needles release a sound like taffeta, and leave the sharp smell of pine on your hand. The DeLacy Creek Trail took me three miles in to a windy lake, through creek meadows awash in* Claytonia lanceolata*—spring beauties—young yarrow leaves, shooting stars, globeflowers, buttercups. New grass rises from the lodged remains of last season, scattered with the kinnikinnick on the forest floor. Water lilies open their bright cups on the slow backwaters of the creek. On Signal Mountain near the Tetons, I spent a morning fearing bears and tromping through forest and high mountain meadows, finding all kinds of things. I* *don't like dense woods, where I can't see far and imagine a bear at every turn, but that's where I saw* Corallorrhiza maculata, *an orchid like a pale flame near a downed tree, off the trail. I walked over to touch, marvel, praise. Up higher, I walked through a sage and balsamroot meadow, to where the horizon opened onto the Tetons, their vast jagged edge, Mount Moran blazing out across the sky, over red gilia, purple delphinium, white yarrow. I sang through berry thickets, and wanted to stay. No one up there anywhere. The high sharp lines of the mountains are imposing, but in the willow flats beneath the Tetons I forgot to look up. Intent close to the ground, I found* Valerian acutiloba *starting with nothing but a guess at a family name.*

*Bear phobia wins out over plants. I passed on the chance to walk to Christian Lake, rich in smell as the trail is, the air humid with willows. The Solitary Walking Woman's Guide in my head adds "miles of willow thicket eight feet high" to my map of places I won't go. But I know I forget this fear sometimes.*

*I forgot bears entirely in Paintbrush Canyon. Nelson walked into some canyon in the Tetons—none of them would have been named on his map, if he had one, in 1899—up as far as the tree line. For me, this trail led up through old-growth spruce and fir trees, with a rich deciduous understory, mosses and lilies near the water seeps, scarlet paintbrush along the trail, and wild clematis draped on the bushes. I stopped at a falls and washed my hands in cold creekwater. The steepest, narrowest part of the canyon, tumbled talus and falling water, was full of the smell of willows and berry bushes hot in the sun—gooseberry, currant, raspberry, black elderberry, western serviceberry, highbush huckleberry. About six miles in, after what seemed like endless noodlings around one canyon wall or the other, Holly Lake lay in a high cirque. Two thousand vertical feet sharpened all shapes, forms, colors, smells, my response to touch—cold water, wind, snow, sun. An orange at tree line is a marvelous delicacy, and the moment I ate it was sweet and broad. Climbers and backpackers go further, but for Nelson and me, the world thins out above the trees. I was content with the lake and the orange. Weather poured over the ragged peaks above the lake, clouds forming and reforming loose fronds and whorls. I heard rocks falling in the canyon, sharp cracks of danger somewhere. And by the time I reached my car and dinner, the same two thousand feet had dulled all this again. But I'll go back.*

*Ready to go home on a foggy morning, and sick of canned beans and cold Ramen noodles, I'm eating a luxurious hot breakfast at Doornan's Chuckwagon, in a huge and ridiculous canvas teepee, coffee, eggs, and bacon next to a bright fire in the floor. I made a point not to look at Nelson's lists this week because I wanted to puzzle over things for myself. Here though fortified with woodsmoke and food, I read:* Fritillaria atropurpurea, Fragaria glauca, Phacelia sericea, Frasera speciosa, *an* Erigeron—*"a discoid form in the hills" he says, like my rayless friend on Bunsen Peak—and a mystery to him too. Along with* Ribes, Gilia, Potentilla, Ranunculus, Castilleja, *on and on. Hundreds of things I know. Seeing the full familiar names in his liquid handwriting makes me cry. I am in a known place. The species names ring like bells out of 1899.*

*It's snowing; weather and fatigue remind me of the pointless knots I left at home. Nelson brought his wife and daughters, acres of women's clothing to keep clean—Alice's job, along with cooking, and pressing and drying plants. I left my daughter Grete with her father's relatives in Montana. My stove is temperamental; I am grateful for someone else's cooking. I eat breakfast with families, two couples with small children. Anyone on this trip with me seems impossible. If I've come to be reinvented in the wilderness, forgive me. I stopped at the Chapel of the Sacred Heart, though, not the Transfiguration.*

HE HAD TAKEN therapeutic trips before, with Alice. In the spring of 1928, thinking a change of scene would do her good, they had gone to Oklahoma and Arkansas, collecting along the way. In early February 1930 he set off for the Southwest alone. He stopped in Fort Collins, Colorado, to visit colleagues and a student he had met at a conference recently who appreciated his help with some plants she was working with for her master's thesis, a flora of Rocky Mountain National Park. Reaching New Mexico on the eighth, and Arizona on the tenth, Aven started looking at things. He listed mesquite, yucca, an oak, something he called "Bear grass" (in quotation marks), and another shrub he left unnamed. And then his list broke into a strange litany, a three-day catalogue of mistletoe. There was mistletoe on juniper in the Santa Fe Hills in New Mexico, on oak in the Rio Grande Hills, on cottonwood in Albuquerque; mistletoe on a green ash tree at Dragoon, Arizona, on mesquite at Wilcox, on oak at Bowie—an oak "common on disintegrated granite," he noted. Even had there been a widespread eruption of mistletoe that year, there were other things to see; alone in his sprawling grief, mistletoe was what he chose to write down. Apart from its long association with the spontaneous Christmas kiss—a first kiss?—the life and body of mistletoe are entirely part of the life and body of the tree it grows on, before it kills its host. After steeping himself in mistletoe, on the evening of 10

February he sat down to write letters, one to his family reassuring them about his health, and one to the student he'd spoken with in Fort Collins, Ruth Elizabeth Ashton. "This strange desert land would be more interesting," he wrote, "if I had an understanding mind with me to discuss it as I drive . . . I want you to know that I have not forgotten you."

Reaching the Baboquivari Mountains almost two weeks later, where he hoped to find *Calhounia nelsonae*, there was another mistletoe on an ash. On 22 February he found what he thought was a different *Calhounia*, along with yet another mistletoe, this one on a juniper. He gave up his search finally, writing to his family, "I did not find the precious plant that I hoped for . . . I tramped the hills for a half a day, each of the days I was there. On the last trip I did find what is probably a new species in the same genus, so that will be one more *Calhounia*. That is some compensation. I may get the other yet." The trip left uncharacteristically vivid and not especially botanical traces in his collection list. He saw a "sprawley viny clammy peren.[nial] herb" in sandy washes in the Fortuna Range, just before he noted "a delicate herb —spreading decumbent" on open plains in the eastern part of Yuma County, neither of which he identified further. He saw a juniper, "decidedly weeping." From Albuquerque on his way home at the end of March, he wrote again to Ruth, arranging to meet her in Fort Collins before he returned to Laramie.

*It's taken almost a week to find my own way into this place, trying and failing to see what Nelson saw in the Red Desert in 1897. There is no way to recover his trip along I-80. The little towns are gone, and I wouldn't be able to stay in homes in the comfortable ranchlands anyway. I was hoping to start near Rawlins and angle northwest into BLM land, beyond the twenty-mile public/private checkerboard along I-80. He was looking for forage plants; he left the most extensive notes on anything he ever saw on Red Desert forage. All that quickly became secondary on this trip.*

*It started in a campground north of Sinclair on the North Platte River. The cement picnic tables were littered with broken glass, the "campground" an empty treeless gravel lot in an isolated place, but I didn't think much about this. Poking around along the wild roses, picking sage, listening to a BLM boundary sign slap a fence post in the wind, by early evening I'd found a few friends and a few strangers, and lay down in the hot car for a nap. There were men and boys fishing downstream. Mayflies filled the air and pelicans fished the* river. Gunfire woke me up—maybe fifty feet from the car, blam, blam, blam, straight into the riverbank. Never mind that this is illegal. I scrambled into the front seat and drove nearly a hundred miles northwest to Lander to start again. The BLM employee I talked to in Lander the next day told me that people dump methamphetamine production trash at that campground sometimes, and blow out the outhouses with guns; a BLM flier notes that among the larks, wild horses, and flowers to see in the Red Desert, you might also see the remains of meth labs. There's a picture. I stay in a BLM campground near Atlantic City and drive into the desert that way. This has meant driving a hundred or two hundred miles a day, to one corner or another and back, past the shredded rubber of other people's unlucky days, back to Lander occasionally, because there's no gas anywhere near Atlantic City.

In the campground, generators run all evening; one visitor has a whole compound set up, with tanks of water, a satellite dish. Razor-faced young men in fatigues wash their arms and chests at the water faucet. The older couple who are the "campground hosts" fish and feed birds, and have graciously offered to take my notes in the early morning saying where I'm going and when I will be back. No one else knows where I am.

I walked up Oregon Buttes near South Pass, a landmark that overlanders used to sight their route. There used to be a jeep trail partway up the east side but it's been closed as a "wilderness study area." The foot trail is not sensible. It goes almost straight up, to the tip of the larger butte, and then a gentler slope nearly along the continental divide to the top, a shallow dome. All of southeastern Wyoming lies on one side, and the Wind River Mountains on the other. A more marked divide is hard to imagine. On the northwest: green land, sagebrush, rolling grass, and through it the road I took; over the divide: ripples and points in gray, pink and white, crenellated buttes, chalky basins and roadless badlands. I waded through sage and what I thought was gorse—something spiny, which turns out to be Atriplex. The wild buckwheat is familiar here, but small. I recognized almost nothing else but sage.

Near Honeycomb Buttes, I parked on the road in the dunes. Prickly pear, prickly Atriplex, prickly everything but sage and rabbitbrush, which are sticky or pungent or both. The earth is laid bare in cracked pavements. The wild onions are almost dried up. Dune grasses wave in slender bunches. There are horse prints in dried mud all over the place, monuments of horse dung along the road. There must be enough for them to eat, I thought, but I couldn't imagine what it was—not just grass, obviously. I couldn't get closer than a couple miles to Honeycomb Buttes because the two-track road washed out *some time ago—every day I try routes with some landmark in mind, and every day I improvise, and walk. There is much checking of map and compass. I won't get "lost," it's impossible not to see where I am, but that has nothing to do with whether the roads on the map actually connect one place to another on the ground.*

*It's easy not to see the desert. Some of the "dunes" on the map are flat, red, baked, broken plazas strewn with great piles of horse manure, and something spiny and unusually green. This is what they eat—* Sarcobatus vermiculatus. *Creasewood. It's a chenopod, like Russian thistle and four-wing saltbush, both of which thrive here. I carried a male and female branch of saltbush around for days before I understood they were the same species. Overwhelmed by it maybe, I finally made myself look at the brush—chenopods, rabbitbrush, and sage. The place is a textbook of chenopods. Though I found conventionally pretty desert bloomers—bee plant and desert plume on the road—really it's the gray-green brush one or two feet high on silty soil stained with alkali that defines this place. The horizontal sublime. I go back to Oregon Buttes every morning.*

*What am I doing here? I wonder about Nelson's biographer's chapter, "From the Red Desert to Yellowstone"—a transition from some obscurity to acknowledgment? This is a challenging place if you're looking for what anything could eat. Why look here? I distinguish one prickly thing from another and fill my car with dust. I am afraid constantly. Amy Wroe Bechtel disappeared in Sinks Canyon near Lander in 1997, but that is a busy place compared to this one—a pretty forest road, a place to picnic. A flier calling for information on her whereabouts, posted inside the campground latrine, has been clawed right through the picture of her face. I am certain her body is out here somewhere. This is a fear I never forget, and "botany" only magnifies it. What I looked forward to, and started in the parks, scatters here in a hundred directions. The only certain thing is that I don't know shit, this week, any week. It is impossible not to think about Mark in this beautiful and difficult place. Nothing will be different when I go home. I dread that, but here is no better. He is certainly angry and may have left town anyway. I try to believe it was no good, whatever I wanted from him, all that time spent wondering.*

*This morning I gave up. I'll see what I see; I can't worry about the rest of it. After visiting Oregon Buttes and finding the road down toward the Pinnacles, I made myself stay put, because there's little point moving. I spent ten hours watching eagles hunt, and the wind blow, picking up a line of cairns stretching southeast on modest rises, mile by mile, marking someone's route silently through the basin. The loose trill of larks is not a question. I saw the horses, two of them, male and female, insouciant and grazing, watching me, circling each other. Their slow dance carried on for hours before they drifted out of sight. The shape of this day resolves into a wide texture of time.*

WHEN AVEN RETURNED to Laramie in March, he lay the *Calhounia* to rest as *Carpochaete Bigelovii*, previously named by Asa Gray. He also arranged an assistantship in the herbarium to be given to Ruth after she graduated in the spring.

Ruth's thesis was the kind of work Aven had hoped to produce in the decades following the publication of his *Manual*. Though he was frustrated by the public's ignorance of botany, and his colleagues' failure to excite ordinary people about their plant neighbors, he put most of this energy into public lectures, and teaching, especially at the science camp. He was a gifted teacher. He was not a gifted popular writer. There is no Yellowstone in his hand like John Muir's. Aven left no extended descriptions of the natural world nor his experience in it. What he wanted to write was a popular flora of the Rocky Mountains, which is a very different kind of document. It's not the kind of thing you'd *read*, "appreciating" some marvelous landscape or other; it is something you'd *do*. It is a set of tools. This was something Ruth would understand.

She graduated from Mount Holyoke College as an English major in 1924, and started working at a camp near Longs Peak in Colorado, teaching nature study. Wanting to homestead and start a girls' summer camp near Estes Park, she bought 240 acres and named it Skyland Ranch. The camp never materialized, but she worked for the park service and started her master's degree in botany at Colorado Agricultural College—now Colorado State University—in Fort Collins in 1925. In a biographical sketch of Ashton, Janet Robertson wrote that in Rocky Mountain National Park, Ruth "stood behind a counter and answered all kinds of questions: 'How far is it to there?' 'Where can you get to—?' and 'What's the name of this flower?' Informally, and without any official recognition, Ruth also gave naturalist talks 'when there wasn't anybody else to do it.'" She enrolled in the Yosemite School of Natural History in California, hoping that her knowledge of Rocky Mountain Park plants and her natural history training would make her a good candidate for a job as a park naturalist. She was disappointed on that count though, and understood her gender to be the problem. Meanwhile, Ruth collected widely in the park for her thesis. With few people nearby who could help her identify plants, she had asked Aven Nelson for help. Park employment was a professional dead end, and she had not yet prepared her flora for publication; no doubt she was grateful for an herbarium appointment in 1930.

Aven's biographer Roger Williams writes, "The herbarium was . . . the scene of a budding romance featuring the seventy-one-year-old curator and his thirty-four-year-old assistant. The family noticed that Miss Ashton was included for picnics and parties with a frequency not enjoyed by previous herbarium assistants, but no one seems to have suspected that a match was in the making." Robertson states simply, "A romance developed." Aven wrote to Ruth that he feared gossip would cast her as an "adventuress," exploiting an old man for his money (or presumably prestige, given her botanical interests). Robertson wrote, "Being married to such a famous botanist opened new doors for Ruth," including her first trip to Europe, to the International Botanical Congress in Prague. But it is possible, as some of Ruth's friends told Robertson, that "she was also stifled by her marriage to Nelson. Only after she was widowed, for example, did she resume long visits to her beloved Skyland Ranch, which Aven found boring because it lacked a large variety of plants." Though no one may have suspected, it's hard to read the field notes and letters from Aven's mourning trip for *Calhounia* without hearing loneliness, intimate companionship, grief, maybe fear, or even Ruth herself on his mind at least, traveling through the desert and corresponding with Ruth, nine months before she came to Laramie to work. We can only guess at Ruth's motivations. Single in her mid-thirties, she married a man with whom she was not likely to have children, safeguarding a career of some kind even if other avenues had already been closed to her. Maybe close male companionship was not something she was willing to give up. They were both brave. A strait-laced sober citizen, Aven faced the ridicule of people he had known in Laramie for decades, and the disapproval of his own daughters (both of whom were older than Ruth); Ruth faced the prospect of caring for Aven at the end of his life in the prime of her own. They married on Ruth's birthday, 29 November 1931, in Santa Fe. And they took a trip.

They recorded their honeymoon as "Collections of Southwestern Trip 1931 —Aven & Ruth Nelson," beginning with #1 on 8 December, prickly pear on a mesa near the state college in New Mexico, and ending with #71, Spanish moss on Highway 90 in Mississippi on 6 January 1932. They collected through New Mexico and north central Texas. She bought some *Ilex opaca*—holly

—in Shreveport on their way to Lake Pontchartrain for Christmas. They were in New Orleans by the new year, where they saw a banana plant at the Court of the Two Sisters. One of them wrote the species name in, playing: *Musa paradisiaca*. They bought "Kumquat Citrus japonica" in a fruit stand. An eaten collection. When they came home to Laramie and began working together as partners, the next notebook picked up their numbered series where they left off for 1932–33, and all the notebooks after that, travel and work, bouquets, and USDA seed collections, woven seamlessly together in the list for the next twenty years.

*I took the tram to the top of Rendezvous Mountain above Teton Village and walked down Granite Canyon. Deposited on the gravelly summit with the parasailers, I walked south through a great bowl, over its edge and down toward the canyon, through a huge meadow rolling away in all directions, full of color and leaves—acres and acres of gilia, larkspur, yampah, gentian, paintbrush. Gilia so bright in bud—orange—so red red in bloom, fading pink to white as they flare. Almost as if they can sustain only so much color for so long. And gilia trading off with paintbrush, as if a meadow can only sustain so much red at all. The trail along the canyon wall skirts steep meadows, the stream fast up high, falling into pools of light. In the hot berry narrows, a downdraft cold as a refrigerator poured off the talus boulders. The old spruce woods were lathered in sun. Fourteen miles down, I could feel every Teton cobble through my boots.*

*On Jackson Lake beneath the Chapel of the Sacred Heart, the yellow balsamroot is done. This place is not timeless. Time moves through the forests, meadows, and canyons, bringing the buckwheat up and pulling the balsamroot down, draining the leaves of color, raising the slender flowering heads of sage—frilly, fragrant, delicate—polishing the new leaves of* Ceanothus, *ripening berries. The slate sky softens pink, red, fuchsia, purple, before the morning is full. On the road, people are on their way to the next place.*

BY THE TIME the Nelsons reached Mount McKinley (now Denali) National Park in the summer of 1939, Ruth had taken their field notes in hand, making most species determinations herself rather than leaving them for her husband. Their collection in the park is entirely in her handwriting, and she wrote their report for the park service. She was a paid expert in a national park for the first time. At the age of eighty, Aven had a taste of another new park, a remote two-million-acre wilderness refuge for Dall sheep and other large mammals, founded in 1917. They were among the first botanists to study it. The eighty-mile gravel road into the park had just been finished by the Civilian Conservation Corps (CCC), and the new hotel opened for the 1939 season. The park's system of carrying visitors into the park by bus was in place as well, a system that still bans private traffic from the road. Beyond the first fourteen miles, the only road into the park has never been paved.

Other collectors provided plants from earlier in the season (they arrived in June), and from areas of the park they did not see. It was not possible to survey the whole park, and Ruth's report acknowledged that any survey was made more difficult by the fact that "so little botanical work has been done . . . in the interior of Alaska." Hazardous terrain and damp weather hampered collection; also, though she didn't mention it in the report, Aven was elderly.

It is easy to read the Nelsons' report as full of extenuating circumstances excusing the limits of their collection, but McKinley officials were under some pressure in the 1930s to make the park accessible to the public, and any effort in that direction was valuable. This included studying and explaining the park landscape to visitors, providing data for management of the park, and physically opening the park so that people could see some of it. The expense of managing so vast an area far from dense settlement, not to mention the continental U.S., had to be justified somehow. The park had become something of an embarrassment to its directors and the federal government.

The park service hired a number of scientists in the 1920s and 1930s to study and publish information about animals, plants, and landscapes of the park, including Adolph and Olaus Murie. The Muries conducted research on wolves, brown bears, and Dall sheep, conducting winter research by dog team, living in cabins throughout the park for many seasons. Ynes Mexia had contributed a collection of plants to the park herbarium; the Nelsons were the first botanists employed specifically to study park flora more systematically, but had only one season to do their work. The park's original two million acres—now expanded to six—have always been trail-less; the landscape includes steep mountainsides, spongy sphagnum moss more than a foot deep, dense willow thickets, unpredictable weather, cold nights, misty chilly days, and treacherous crossings over braided gravel riverbeds when the water rises, which is frequently. Add to this the constant danger of running into brown bears and moose—the question is not *if* they cross your path but *when*—and you have as challenging a scene as imaginable for any thorough study, especially if bears and moose are not the point.

Their Alaska trip was squarely in line with Ruth's interest in popular flora of national parks. Aven had spent the summer of 1899 in Yellowstone, but that season provided material for his Rocky Mountain botany and the herbarium collection (as well as private memories for himself and his family, which he never recorded). The Yellowstone trip was never conceived as research for a field guide. This was the purpose of their study in McKinley. The Nelsons succeeded in getting over seven hundred plant specimens, but Ruth understood that someone would have to spend another season or two "before an accurate ecological study of this area could be prepared," or "before a popular bulletin such as one of those of the series published by the Park Service on the plants of the National Parks, (Rocky, Glacier, Yellowstone), could be satisfactorily produced." Of course, the field guide to Rocky Mountain National Park was her own. "Our purpose in making these lists," Ruth wrote, "is to give a rough sketch of the floral aspect of each of those types [of habitat] by naming some of the commonest or most conspicuous species most likely to be found in each."

The Nelsons were looking for exactly those plants likely to be featured in a field guide: common and conspicuous—as opposed to rare and inconspicuous, the province of experts—indicating well-defined habitats. This was the formula for her other field guides. Ruth also provided dozens of photographs she had taken in Alaska herself for the report, which might have illustrated a field guide. She never produced one for McKinley, but this isn't surprising. Had Aven been younger, they could conceivably have returned. But the Muries worked for years to produce their work on the park, and were willing to live there; Aven and Ruth had strong connections to Wyoming, Colorado, and the Southwest.

*It feels like Wyoming, only with smaller trees and more water. Neither goes up or down very far. Darkness never falls; we subsist on three or four surreal hours of sleep at night. Grete is a little uncertain of the whole enterprise here, especially after an all-night rain and all-day drizzle tested the tent early on. But she brightens with a black wolf, caribou running along the braided river gravel, running on the road, great mother brown bears padding across arctic prairie with their cubs, or asleep yards from the bus windows. Her grandfather just returned from Alaska as we got here. He'd gone with her father and uncle. This was Bud's only vacation, his last trip. I am glad she is here. At the Toklat rest stop, she walks across the river rocks, arms out wide, smiling. I took a picture. Listening to conversations in several languages at the hotel, we've had our fill of hotel hamburgers, before the ranger talks at the auditorium in the evenings.*

*Every night, the naturalists begin their talks with the same warning and physical pantomime about moose and bears. If you meet a moose, back off. But for the brown bear, there is a whole series of possibilities, each grimmer than the last. If the bear hasn't seen you, go away. If the bear has seen you, make yourself big, arms over your head—but don't just stand there that way, wave your "antlers" and talk, "hey bear, hey bear," so it knows you're a human being. If the bear charges, this is probably a bluff charge, so don't run. Pepper spray may or may not help you here. Keep talking to the bear. If the bear charges for real, curl up, hands clasped behind your neck, play dead. The park has done everything it can to keep bears separated from people and their food. But it isn't just human food that's important. Park employees tell people to keep quiet around all wildlife, but especially bears. No one wants the bears to become too familiar with the human voice.*

*The tour drivers remind people repeatedly that this is a wilderness. Either you wouldn't know, or you have to be told again, it's not scenery. Unless you're out there with your bear canister for food, and waterproof gear, obligated not to walk single file to keep from leaving a trail, you're on the bus with binoculars, and shutters snapping. Species-segregated. Our space and bear space. Caribou, coyotes, and wolves have road privileges, of course. Ptarmigans flap and nudge their chicks in the gravelly roadsides.*

*We couldn't camp at Wonder Lake, where the Nelsons' 1939 Christmas photograph was taken—the campground was full all week. Two other campgrounds are closed indefinitely because of wolf dens nearby. We've been on the buses, with hundreds of other people. The campground at the train station is packed every night. Everyone stores their food in the same big bear locker, cooks and eats at the same hours in the same dusty plaza between the split log benches, brushes their teeth at the same faucet, flushes the same blue stuff down the latrine. There's a shelf in the locker for food left by departing visitors for the rest of us—I'm eating a gift of dry apricots. I'd finished my own. I'll leave macaroni and cheese. Looking for a solitary place to write, I have a bench at the depot, which is deserted. Spirit of Nenana, Spirit of Palmer, pull big tanker cars along the tracks toward Anchorage.*

*Taiga forest, alder hillsides, a rhythm of mist and sun: we've walked everywhere we can near the campground, up Mount Healy, over to Horseshoe Lake, to look for things. There is a lot of walking, from the campground to anywhere. We know the network of paths in the trees pretty well. Grete finds new routes all the time. A guidebook with lots of photos lists plants at the rest stops, which has been perfect for me; on each ride, sixty miles, eighty miles, twenty miles into the park, that's where we stopped. I rummaged through the Jacob's ladder and* Rydbergia *with whatever time I had.*

*Rain and fog wrapped the world close for days, so we could see only part of a valley, a high ridge, a meadow, at a time. The mountains we could see were as tall as the Snowies at home, really foothills here. On the eleven-hour drive to Wonder Lake and back, the bus windows were foggy. At the Eielson ranger station some sixty miles in, marks on the floor gauged to your height show you where to stand to look at an outline painted on the big window—it's the huge outline Denali would fill there if the sky were clear. It wasn't. We stopped at the end of the road to walk to the lake, hats and hoods drawn around our faces because of the mosquitoes. As people snapped their cameras, one of our companions on the bus offered to take a picture of the two of us with Grete's camera, and we returned the favor. I washed my face in lake water. It was hard to imagine the Nelsons there—no trees in their picture on the shore, and it was bright. We saw acres of black spruce.*

*A solitary biker we recognized from the campground sloshed his way in the cold rain down the muddy road, sometimes ahead of us (the buses stop all the time to pass each other, and so we can watch animals), sometimes behind us. I worried about him and those many miles until I saw him washing at the faucet later.*

*Our last day on the bus was sunny. At Highway Pass, Denali spread a mass of brightness over yellow arnicas. It is the biggest thing I have ever seen. Or never seen, had we left a day earlier. It is vast and permanent whether anyone sees it or not.*

*I'm not impatient that we couldn't walk into all this. Days have been full with laughing, looking, housekeeping at "squirrel camp," pretending to fight the stove and giving in to warm cafeteria food, and the occasional soda from the little store on the road. It's even fine that we ran out of cash days ago. Our tent is full of books—on bears and wolves, plants and the park, postcards, gifts. Of course, we have to carry all this out to the train tomorrow on our backs. I won't even look at Ruth's notes until we get home. Partly I don't want to know the depth of my "scholarly failure," not yet, not here. I am too happy. Really I don't give a damn. I'm ready to go home.*

THEY MADE THE most of their time. Based at the CCC camp, now called C-Camp, near park headquarters, they recorded plants found at the rest stops—Teklanika, Eielson, Toklat—along the park road. They collected at the campgrounds, including Savage River, Riley Creek near headquarters, and Wonder Lake at the eighty-two-mile mark at the end of the road. Ruth did get off the road to collect at Muldrow Glacier, a massive active glacier and prime scenic attraction, but as part of a pack trip, an excursion available to any visitor for fifteen dollars a day. The Nelsons were working tourists whose work and play fit neatly into their assignment for the park service. Aven had collected in culturally interesting or important places, including Yellowstone, but also Frijoles Canyon in New Mexico, part of which became Bandelier National Monument in 1935. This is an Anasazi ruins site, though he didn't mention the ruins in his 1930 field notes. He had never recorded those places for a popular audience. Ruth was developing her life's work branching out to other parks, and Aven ventured into botany of popular interest for the first time. No one has produced a very extensive guide to McKinley plants. The best available, like Ruth's notes, includes photographs rather than botanical keys, and checklists for plants at the rest stops, from which most of the park's visitors experience the place. It expects little effort on the part of the reader, whose time with the plants is very likely limited to the fifteen minutes or so allowed at each scheduled stop.

*He couldn't come with me, I couldn't come home to him. So far off, even when I could smell the sun on his skin. He said his soul isn't here anymore—here, where? I couldn't say anything, I could hardly breathe. I don't think I've ever heard such close words that opened such an empty place. I will never understand who he is in this world.*

AFTER HER FLORA of Rocky Mountain National Park, Ruth's second publication was *Wild Flowers of Wyoming* (1936), an illustrated field guide that filled a need for popular plant identification. Aven had been answering questions for years about Wyoming wildflowers, regretting that an early circular had not been republished, and recommending his own *Manual*. Though it remained a standard botanical reference for Wyoming and the Rocky Mountain region for decades, Aven's *Manual* was neither illustrated nor especially easy to use for a layperson. Ruth's booklet was, first of all, about "wildflowers," not a "manual of plants," obviously written for a popular audience who had little or no botanical training but who wanted to go outside and know something about the plants they saw. This remained in print a long time, with new editions in 1962, 1968, and 1984 (the last updated by Rocky Mountain Herbarium director Ron Hartman). Ruth completed a third field guide, *Mountain Wild Flowers of Colorado and Adjacent Areas* in 1967, and another guide specifically for a national park, Zion, in 1976 (which remains in print). Each one of these books, and even the report on Alaska, represents months of collecting, and more months checking the identities of plants and updating their names, and refining the keys to the plants repeatedly just to maintain an existing guide in print.

Ruth's field guides are botanically accurate introductions, in some ways "simplifications" of her science: she explains the difference between common and scientific names, the vocabulary of plant anatomy, and the process by which you identify a plant using the keys—basically either/or questions about the size, hairiness, color, form, shape of the seed, and any number of other characteristics that lead you successively to a division, a family, a genus, and a species for whatever floral mystery you have in your hand. She had the foresight to include illustrations, occasionally color photographs, encouraging the genuine novice with some hope of seeing exactly what was pictured. But these are not splashy picture books to look at. With patience, her keys are in fact more useful than any illustrations. And if a popular field guide is "simplified" botany, the process of identification is anything but a simple experience.

Written in language more everyday than Latin where possible (with a glossary just in case), the keys are precise and lead to many more species than the number of illustrations that could be included. She doesn't attempt to be comprehensive; what she calls "difficult" genera or species are left at a distance. You may not get much closer than a genus name for certain erigerons, asters, astralaguses, or penstemons, but then you won't be overwhelmed with them either. Still, the keys are seductive; they pull the reader into botany, perhaps unexpectedly, by having them learn to use and rely on the system of distinctions—learning along the way what to expect in certain families or genera, eventually "teaching" the reader enough about plant forms (by repeated experience) to be able to start with a family rather than a division, or a genus rather than a family. Left at the end of a trail amid the penstemons, beautiful and variable flowers in blue, purple, red, and white, with their five stamens and little bearded tongues, it is not hard to want to know more, beyond the veil of "difficulty" that drops before the penstemon at your feet is known with certainty. And the plant itself has become seductive. An illustration makes the easiest comparison possible, but using a key means looking much closer: handling the plant, testing it for stickiness, smelling its flowers and leaves, noticing how the leaves attach to the stem, how the petals of the flower are joined or divided or have a bright yellow smudge in the center, looking under the petals at the fringe of sepals holding them, and comparing the shape of the leaves near the ground with those at the top of the stem. You spend some time with it and know the plant sensually—one might say empirically—in a way you are not likely to forget. What you learn is much more than its name. You learn how to look, distinguish, compare, anticipate. Useful skills, through and beyond plants. Ruth's gift as a field guide lay in part in her ability to draw you to the plants she knew, and lead you to your own recognition of them firsthand, believing this was possible. At the heart of the process is an encounter that is not strictly botanical.

Ruth's writing for popular gardening and nature magazines began at the very end of Aven's life, when he was living in a nursing home. He died in 1952. She may have felt she couldn't pursue popular writing before then, or maybe she simply wasn't ready. She was a regular, but not voluminous writer for the popular press. These pieces speak to her audience's love for domesticated nature without leading them so purposefully through a botanical exercise. She wrote about the birds she was likely to see in her yard and the plants in bloom there in different seasons, the wild plants she recommended transplanting to new suburban gardens on the Front Range in Colorado, and the success of xeriscaping. She was an avid gardener. Her writing was clear, observant, and vividly descriptive, but readers of this kind of writing are often consumers of images who let other people know nature well and translate it for their reading pleasure. Ruth's field guides represent another kind of effort altogether—on her part and on the reader's. The audience of her field guides would come closer to knowing something of her own relationship with the world through plants, by following her movements and observations. She learned what she knew by experience, and she asked for help. "Nature" is only part, and maybe a small part, of the process.

Ruth's *Handbook of Rocky Mountain Plants,* first published in 1969 and revised in 1992 by Aven's biographer, Roger Williams, was her most extensive field guide, and her own favorite. True to her interests, the book offers a map of national parks and monuments located within the region she covers, understanding that readers are most likely to see and investigate plants on visits to these special places. She dedicated the book to Aven, "Inspiring Teacher, Faithful Friend, Beloved Companion." She perfected her craft as a botanist with the public in mind, shaping the end of her husband's career, inhabiting their notes as a protégé, and as a partner. Her *Handbook* fulfilled his wish to see a popular manual of plants of the region he loved and studied for so long. It was suitable as a book "presented to plant lovers of the Rocky Mountain region as a memorial" to her husband. The book is both the completion of Aven's lifework, and a cornerstone of her own.

*When I met Roger Williams in the herbarium one day, he looked over a page of Aven Nelson's early collecting and rattled off the list of genus names—* Erigeron, Aster, Salix . . . *"Very common," he said, and went about his own search for* Oreocarya. *Roger conjures an entire landscape, and his full memory, from a glimpse at these names. Mine is different from his, but no less familiar. Or common. He returns to botany too. My copy of Ruth's (and Roger's) handbook is battered, with a new cover, and full of scraps from Laramie to Fairbanks. "Happy Days!" Roger wrote to me on the title page.*

Aven and Celia Alice Nelson, 1920s. Courtesy American Heritage Center, University of Wyoming. Aven and Ruth Nelson Papers.

Aven Nelson working with specimens in his tent in Yellowstone National Park, summer 1899. Courtesy American Heritage Center, University of Wyoming. Aven and Ruth Nelson Papers.

Ruth and Aven Nelson in front of their tent, with their vasculum and specimens, at Wonder Lake, Mount McKinley National Park, July 1939. The Nelsons used this photo on their 1939 Christmas cards. Courtesy American Heritage Center, University of Wyoming. Aven and Ruth Nelson Papers.

Many blue penstemons
Goldenrod
Geranium
Yarrow
Yampah
Cow parsnip
Yellow and white sweet
clover
Harebell
Lupine
Fireweed
No surprises
Pedicularis parryi var.
purpurea
White oxytropis
Agoseris glauca
Heartleaf arnica
Polemonium viscosum
Polemonium pulcherrimum
Cerastium arvense
Mertensia oblongifolia
Cirsium scariosum
Frasera speciosa
Polygonum bistortoides
Aster engelmannii
Sedum lanceolatum
Small solidago
Smell of water

Musk ox
Monkshood
Alder
Dogwood
Larkspur
Juniper
Carrot
Roses
Bluebells
Fireweed
Feltleaf willow
Diamond willow
Eskimo potato
Shrubby cinquefoil
Black wolf
Pink castilleja
Solidago multiradiata
Gentiana prostrata
Myosotis alpestris
Mertensia paniculata
Geranium erianthum
Campanula lasiocarpa
Aconitum delphinifolium
Delphinium glaucum
Polemonium acutiforum
Aster sibiricus
Brown bear

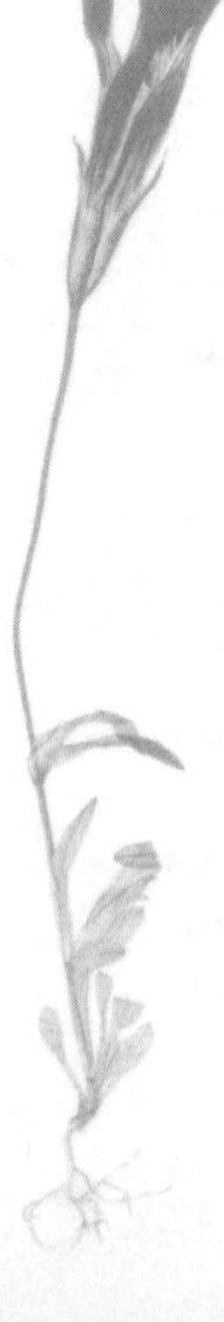

# Red Desert Reprise

A year later I go back. I go back to walk up Oregon Buttes again—a palpable being on the edge of the desert. It is as much a sound as it is anything else. I go back to see someone. Some of his country is on the other side of the divide, in the mountains, with the elk, moose, deer, creeks, fish, and hunting camps; drawn figures on rock, dogbane under a fringe of butterflies, roses, the press of sage and aspen on calves and shoulders like a human touch. Walking in the desert, maybe he is humoring me.

Parts of him are here, too. We startle desert elk, rich red animals. Great spreading crowns of antlers rise up out of sagebrush like landforms. Sage grouse clatter wings in a spray of flight. I brought the dogs who nose and trail through a dense world of smell, weaving dog, sage, bird, and mammal together in sinuous travel.

This time, the desert is different.

Its lines and seams are touched by water and we walk toward springs. He is looking for ancient camps. I find irises, fat green grass in a cloister of tall aspen. He finds shards of red, pink, and gray stone, chips struck by a blow and fallen from old work exactly here. We see the round roof of a sheepherder's wagon far off, and a signature on a wall of an abandoned house: Abraham Perez 1970. There are people of many kinds with us, a very long past.

Show me your favorite place, he says, and we drive for a half hour or so to a crackled alkali plane interrupted only by greasewood, several acres broad. It is deceptively simple. The drought-patterned ground gives way a little under our feet and the dogs bound off through the soil along a line of horse prints. Greasewood flares green on mounds of collected earth. There is texture everywhere, and everywhere repeated. Full cadences of horse, earth, sun, branch, leaf, lines of dry and vanished water, new pressed prints. Lying on my back in the middle of it, I can hear the dogs' tags, and I know he is walking through the rim of this shallow bowl; when I meet him, he's got a handful of chips. There are dozens within a few minutes' walk. I don't know how to see them; he lifts them in his palm for me, and leaves them. I'm drawn to white rocks every time, but within an hour they appear: a smooth fleck of gray, a sharp shiny piece of red. Next to the rust lichen on the bark of greasewood, leaning into sand with the buckwheat.

It is the same place. It is not the same place.

# Notes

WORK IN PLACE

1. Primary sources for this chapter are Aven Nelson's papers and the records of the Departments of Botany and Geology at the University of Wyoming, available at the American Heritage Center, University of Wyoming, Laramie, Wyoming. Historians of science at Oregon State University suggested questions for an earlier version of this essay in March 2001. American Heritage Center material is cited by an abbreviated name of the collection, with box and folder numbers. The collections I used are abbreviated as follows: ANP: Aven and Ruth Ashton Nelson Papers, University of Wyoming American Heritage Center, Accession #400013; BR: University of Wyoming Department of Botany Records, Accession #545001; GR: University of Wyoming Department of Geology and Geophysics Records, Accession #545004.
2. See Henrika Kuklick and Robert Kohler, eds., *Science in the Field.*
3. Harold Dorn, *The Geography of Science*, xi.
4. Joseph Ewan, *Rocky Mountain Naturalists*, 271.
5. Bassett Maguire, "Highlights of Botanical Exploration in the New World," 218.
6. Roger Williams, *Aven Nelson of Wyoming*. Ruth Ashton Nelson published a nonspecialists' *Handbook of Rocky Mountain Plants* (1969), which fulfilled Aven's wish to see a popular manual of his region, and was in many ways the completion of his life's work as well as a cornerstone of her own. It is still in

print, revised and edited by Roger Williams (1992). For historical facts about the University of Wyoming, I used Wilson Clough, *A History of the University of Wyoming 1887–1964*, in addition to my own familiarity with the university. For historical and biographical facts about Nelson in the account that follows, I relied on Williams's exhaustively documented *Aven Nelson*, in addition to primary source material (that Ruth Nelson made available to Williams for the biography, and thereafter became the Aven Nelson Papers at the University of Wyoming). The facts of Nelson's life are as Williams presents them. I don't, however, share Williams's interpretation of Nelson's accomplishments. For Williams, former professor of history at the University of Wyoming and amateur botanist himself, "the life of Aven Nelson was extraordinary in many ways" (ix). Williams presents his subject as a significant player on a large scale wherever possible: "a truly seminal teacher . . . the preeminent botanist in the Rocky Mountain region . . . the patron saint of Wyoming rural folk," who "built the Rocky Mountain Herbarium from scratch and gave it the impetus to become what it is today, the foremost collection of plants between St. Louis, Austin, and the West Coast" (ix). My emphasis is different. Rather than transcending his education and professional appointment, Nelson's ordinary rather than illustrious experience over decades in a specific place reveals the local forces that shaped his work.

7. Aven Nelson, "Discipleship," address given as past-president to the Botanical Society of America, 30 December 1935. Typescript, ANP, box 11, folder 2, 2–3.
8. Williams, *Aven Nelson of Wyoming*, 17–18.
9. Ibid., 21–22.
10. Ibid., 24–25.
11. Aven Nelson, "Discipleship," 11.
12. Williams, *Aven Nelson of Wyoming*, 30.
13. For Robinson's condescension on this point, and Nelson's ongoing predicament sending specimens to various authorities for determination, see Williams, *Aven Nelson of Wyoming*, chapter II, "A Faculty-of-All-Trades," 17–43.
14. See Williams, *Aven Nelson of Wyoming*, chapter III, "From the Red Desert to Yellowstone," 44–64, and chapter IV, "From New York to Chicago," 89–111. Nelson and Rydberg participated in a controversy about the emerging standards of American botanical nomenclature. Nelson needed influential patrons to get his plants published, and was caught between different "schools," literally. His field knowledge but also his eventual commitment to "lumping" rather than "splitting" prompted John Coulter's choice of Nelson to revise his 1885 manual.
15. Aven Nelson, "Wyoming Junipers," *Botanical Gazette* 25, no. 3 (March 1898): 196–98 and "The Rocky Mountain Species of *Thermopsis*," *Botanical Gazette* 25, no. 4 (April 1898): 275–78. Nelson began a series on "New Plants From Wyoming," in the *Bulletin of the Torrey Botanical Club* in 1898 as well, which continued through 1904.
16. John M. Coulter, *Manual of the Botany (Phaenogamia and Pteridophyta) of the Rocky Mountain Region, from New Mexico to the British Boundary*; Aven Nel-

son, *New Manual of Botany of the Central Rocky Mountains (Vascular Plants)*. Nelson shared authorship with Coulter as long as the *Manual* was in print, though his revision was thorough. A manual lists plants in phylogenetic order with textual description of them in Latinate botanical English. It is not usually illustrated.

17. Williams, *Aven Nelson of Wyoming*, 123–26.
18. See for example Merritt Lyndon Fernald's preface to Gray's *Manual of Botany*, v.
19. Elizabeth Keeney documents the widening distance between professional and popular botany, including women's marginalization in this field, in *The Botanizers*. Marcia Myers Bonta describes women's overlooked field contributions in *Women in the Field*. Margaret Rossiter does the same for science in general in *Women Scientists in America*. Asa Gray set the standard for botanical teaching and professional practice through the turn of the century and beyond. See his *Lessons in Botany and Vegetable Physiology* and *Introduction to Structural and Systematic Botany and Vegetable Physiology*. Representative textbooks of Gray's outlook around the turn of the twentieth century include William Chase Stevens, *Introduction to Botany* and William F. Ganong, *The Teaching Botanist: A Manual of Information Upon Botanical Instruction*. Ganong, one of Nelson's Harvard instructors (and professor of botany at Smith College), was a kind of Boston Gradgrind (as in Dickens's Gradgrind), emphatically eschewing sentiment in favor of scientific rigor, which was probably an asset to his Smith College students who otherwise might have been dismissed out of hand. Charles Goodale, another of Nelson's instructors, noted in his lectures that botany was "Formerly a school girl's study" whose object was "the preparation of albums" ("Teaching of Botany," 1891–92, manuscript lecture notes, BR, box 29, folder 3). Nelson's career embodied both sides of the widening divide between scientific and popular (sometimes sentimental) botany.
20. See for example William J. Hooker, "On the Botany of America."
21. Keeney, *The Botanizers*, 127.
22. Reed C. Rollins, "Taxonomy of Higher Plants," 192–93.
23. Nelson regarded Rollins as among his most exceptional students. Rollins wrote to Roger Williams in 1979 that, when the American Society of Plant Taxonomists formed in 1936, Nelson felt it was so important that the young man become a charter member of the organization, he paid for Rollins's membership himself —three years after Rollins had graduated from the University of Wyoming. See Williams, *Aven Nelson of Wyoming*, 251, 301.
24. See Keeney, *The Botanizers*, chapter 10, "The Nature-Study Movement: The Legacy of Amateur Botany," 135–45.
25. Liberty Hyde Bailey, *The Nature Study Idea*, 5.
26. Aven Nelson, examination questions, Nature Study, 25 March 1905, BR, box 28, folder 19.
27. Keeney, *The Botanizers*, 140.
28. Ibid., 138.
29. Ibid., 139.

30. Aven Nelson, "The Rocky Mountain Herbarium," *Phi Kappa Phi Journal* 23, no. 4 (December 1943): 135–40. ANP, box 23, folder 3.
31. Keeney, *The Botanizers*, 138.
32. See A. Hunter Dupree, *Asa Gray*, 336.
33. William Whitman Bailey, *The Botanical Collector's Handbook*, 1.
34. Ibid.
35. Ibid., 2.
36. Ibid., 2–4.
37. Aven Nelson, "The Rocky Mountain Herbarium," 136.
38. Theological overtones and references are woven throughout Nelson's incidental professional addresses, including even the title of "Discipleship." His sermons expressed his eagerness to teach evolutionary theory in reverence for the Creator, including "Design in Nature in the Light of the Theory of Evolution," manuscript, 1893, ANP, box 11, folder 1 and "The Struggle for Existence" typescript, 4 November 1900, ANP box 23, folder 5.
39. Every profession faces this problem. In botany, Daniel Culross Peattie bemoaned his erstwhile colleagues' neglect of and disdain for the intelligent reading public, particularly women, in his essay, "On the Popularization of Botany." Botany did not face a simple problem of "popularization"—it's not as if botany were a new science or practice—but it had become newly arcane in an era when the public remained very interested in many aspects of the natural world, sentimentally, recreationally, as well as intellectually.
40. Aven Nelson, "Science and the Laity," "Science and the State," typescript and manuscript addresses, no date, ANP, box 12, folder 9.
41. Quotations from Nelson in this and the preceding paragraph are from "Science and the Laity," typescript, no date, ANP, box 12, folder 9.
42. Quotations from Nelson in this paragraph are from "Science and the Laity," manuscript, no date, ANP, box 12, folder 9. The text of this draft is entirely different from his address for Ramaley; its rhetorical extravagance, somewhat beyond Nelson's usual style, may indicate it was a private rant.
43. Nelson, "The Dual Purpose Manual," typescript book prospectus, no date (evidently 1926; see Williams, *Aven Nelson of Wyoming*, 229), ANP, box 11, folder 4, 6.
44. The spirit of Nelson's proposed revision was captured in Ruth Ashton Nelson's *Handbook of Rocky Mountain Plants*. Drawn to the public, Aven was a professional with some institutional recognition who struggled to remind his colleagues of the needs of the public; drawn to scientific botany but unlikely to receive professional recognition, Ruth created a professional identity through field guides Aven would or could never write. In the time they spent traveling and working together, these paths coincided substantially.
45. Corabelle Ewel to Aven Nelson, February 1929, and reply, BR, box 22, folder 1; Tacetta Williams to Aven Nelson, 16 August 1929, Rose Snell to Aven Nelson, 27 July 1929, and replies, BR box 22, folder 2. See also letters from and to Mrs. A. D. Williams and Mrs. Irvine Rendle in BR, box 22, folder 2.

46. Mrs. Charles Bigelow to Aven Nelson, 17 August 1929, and reply, BR, box 22, folder 1.
47. Mrs. Allen to Aven Nelson, 27 May 1929, and reply, BR box 22, folder 3.
48. "Nelson's Lectures on Wyoming Flora Draw Large Crowds," newspaper clipping, no date, ANP, box 23, folder 18.
49. There are no towns between Laramie and Centennial; a passenger rail line connected these places that, in local experience, are part of the same landscape. The Snowy Range Mountains form the western edge of the Laramie Basin.
50. Aven Nelson to J. Arthur Harris, professor of botany at the University of Minnesota, 9 March 1929, BR, box 22, folder 1.
51. See Frederick W. Reckling and JoAnn B. Reckling, *Samuel Howell "Doc" Knight*, especially chapter 2, "Science Camp" (45–57). Science camp records are scattered and thin. The Recklings' book is based on Knight's published geological work and students' memories of Knight as a teacher, including their stories about the science camp. A potentially useful nomination of the science camp buildings to the National Historic Register is missing from Wyoming State Historic Preservation offices. No information is publicly available about the Department of Zoology's participation in the camp. The camp was closed in the 1980s and the buildings sold for a small sum (the property is leased from the forest service); the lodge and cabins remained standing, operated as the somewhat shabby but locally patronized Snowy Mountain Lodge, recently refurbished as the "Triple Rhino Eatin' Place," a resort catering to out-of-state snowmobilers.
52. See registration lists of science camp participants (which often included students' home institutions), 1923–42, GR, box 41, folder 4.
53. "Class Expresses Appreciation of Dr. Nelson's Work," *Branding Iron*, 13 August 1924, 1–4. ANP, box 23, folder 15.
54. "Botany Classes on a Study Jaunt," *Branding Iron*, 20 July 1926, 3. ANP, box 23, folder 15.
55. Iris Harrington, "The Old Botanist," *Branding Iron*, 20 July 1926, 2. ANP, box 23, folder 15. Harrington was no schoolgirl. Mother of a young man who would graduate in 1927, she was completing graduate work in botany, and was principal of the Indian School at Albuquerque, New Mexico.
56. Ihsan A. Al-shehbaz, "Reed Clark Rollins December 7, 1911–April 28, 1998." Rollins established a fund for supporting field research at Harvard as his legacy.
57. Aven Nelson, "Centennial Valley and Adjacent Hills Botanists' Paradise," *Centennial Post*, 14 June 1913. ANP, box 23, folder 14.
58. Anne Carole Moore, "From Medicine Bow to University," *New York Herald Tribune Books*, 31 July 1927, VI, 6. ANP, box 23, folder 16.
59. Agnes Wright Spring, "Botany Bank: 50-Year-Old Herbarium at University of Wyoming Is World Famous," *Rocky Mountain Empire Magazine*, 30 November 1947, 6. ANP, box 23, folder 17.
60. See Ray Allen Billington, *Frederick Jackson Turner*.

1. See the *Oxford English Dictionary* entry for "herbarium," and William T. Stearn, Introduction to Linnaeus's *Species Plantarum*, 103–4.
2. David Winter, "Preface," *The Pressed Plant*, 9.
3. Asa Gray noted in 1857 that 16.5" x 10.5" or 11.5" was "an approved size." Linnaeus used herbarium sheets of 11" x 7" in the eighteenth century. Herbarium sheet size forces the hand of the collector to arrange the plants in suitable shape, rather than the other way around, a subtle but telling emphasis. See Gray, *Lessons in Botany and Vegetable Physiology*, 201.
4. University of Wyoming Rocky Mountain Herbarium guideline flier, "Collection, Preparation, and Preservation of Herbarium Specimens," 1997. See also Richard Klein and Deana Klein, *Research Methods in Plant Science*, 80.
5. Klein and Klein, *Research Methods in Plant Science*, 80.
6. Bailey, *Botanical Collector's Handbook*, 94–95.
7. Pliny The Elder, *Historia Naturalis*, quoted in Victoria Dickenson, *Drawn From Life*, 83.
8. Agnes Arber, *Herbals*; Dickenson, *Drawn from Life*, 82.
9. For histories of botanical illustration in print reproduction, see Arber, *Herbals*, and David Bridson, Donald Wendel, and James White, *Printmaking in the Service of Botany*. See also Handasyde Buchanan, *Nature into Art*.
10. Dickenson, *Drawn from Life*, chapter 3, "The Living Image," 68–104.
11. Ibid., 175.
12. Arber, *Herbals*, 139–40, 142. See also Stearn, Introduction to Linnaeus's *Species Plantarum*, 103–4.
13. Stearn, 103.
14. See Winter and DiNoto, *The Pressed Plant*, 28–31, 80–89, for information on this practice and especially photographs of specimens from bound volumes produced in 1792 and 1816, the latter in an edition of one thousand copies.
15. This discussion is based on Dickenson, who cites Arber, *Herbals*, Brian Ford, *Images of Science: A History of Scientific Illustration* (London: The British Library, 1992), and Elizabeth Eisenstein, *The Printing Press as an Agent of Change: Communication and Cultural Transformations in Early Modern Europe* (Cambridge: Cambridge University Press, 1979).
16. Wilma George, *Animals and Maps*, 23. On the use of animal emblems in particular, see also William Ashworth Jr., "Emblematic Natural History of the Renaissance."
17. Dickenson, *Drawn from Life*, 34.
18. Ibid., 82.
19. Arber, *Herbals*, illustrations: 15, plate v, 248, 267; Apuleuis' ancient advice reprinted in the fifteenth century: 39–40. Broad-leaved plantain was used widely by indigenous people in North America to treat snakebite as well when this European plant became naturalized, sometimes in advance of the human immigrants. Arber was and is not alone in dismissing the web of superstition sur-

rounding the mandrake, but it does not seem so extraordinary that a root with psychoactive properties (from inducing sleep and pain relief to delirium, with grisly associations at least since the Romans used it to prepare their victims for crucifixion) would be treated with elaborate symbolic formality, and its most immediate application—altering the experience of human being—would be indicated somehow in its illustrated icon.

20. Arber, *Herbals*, 146, 191, 193.
21. Dickenson, *Drawn from Life*, 82. Again, Dickenson relies on Ford, *Images of Science* and Arber, *Herbals*.
22. Frances Yates, *The Art of Memory*, 104.
23. See Yates, *The Art of Memory*, 101–4. Why twenty-two *trionfi* and not twenty-one? Perhaps because the tarot trumps include a "zero," the April Fool. This may be an artifact of the development of medieval and Renaissance mathematics as much as the history of social and psychological symbolism.

There is a frustrating lack of serious historical attention to such systems of organization and interpretation, and practitioners of these alternative analytic frameworks are not necessarily careful historians in print. Looking for explicit connections between astrology or the tarot and memory mansions, I found this: "From *The Mind of a Mnemonist*, written by A. R. Luria, and translated from the Russian by Lynn Solotaroff (N.Y.: Discus Books, 1969), there is a theory that Tarot may have been a mnemonic device for monks; perhaps a visual filing system to remember all that they were to record in their manuscripts" (Angeles Arrien, *The Tarot Handbook*, 17). Whose theory? Perhaps an unnamed editor of the edition Arrien found. Nowhere does Aleksandr Luria's own text mention the tarot, but the book is instructive anyway.

Luria was a Russian psychotherapist who studied a client early in the twentieth century for what appears to have been a debilitating overdevelopment of traditional mnemonic virtuosity. The client, "S.," was fluent in Yiddish and Russian, and grew up in a household where his father remembered every book in his bookshop by its place. Imagined figures swarmed over S.'s prodigious feats of memorization—often lists or series of nonsense syllables, performances by which he made his living—which hinged on his ability to remember routes through whole cities that he recalled with remarkable detail. He was evidently a master of the disparaged technique of rote memory, and the narratives he created with it were infinitely dense. S. could not understand written texts easily; even conventional figures of speech were captivating cul de sacs, and the very syllables of a word (probably typography as well) had colors or associated sounds for him, all of which transported him quickly off the page. Luria didn't ask about the historical and filial sources of S.'s impressive technique, or his enormous fund of images and associations, but his client was clearly following traditional rules for images and their placement. S. appears to have been a direct, recent heir of the old memory arts that included at least fragments of the astrological system, the Cabbala, and likely (but not certainly) the tarot. See Luria, *The Mind of a Mnemonist: A Little Book about a Vast Memory*, translated by Lynn Soloratoff.

I discussed possible connections between the development of the Renaissance tarot and formal memory technique of the same period with Marcus Jungkuth, a Jungian psychologist, and secretary general of the Ordo Templis Orientis (the Hermetic Order of the Golden Dawn, which was tarot master Aleister Crowley's scholarly and spiritual home early in the twentieth century). Jungkuth wrote to me that the historical origins of the tarot are too murky and too ancient to conclude that the system was developed deliberately as a mnemonic structure, but that the tarot has certainly long been useful in that capacity. Personal communication, August 2003.

24. Arber, *Herbals* (164, 166) notes that "the earliest [herbals] show scarcely any trace of natural grouping, the plants being, as a rule, arranged alphabetically . . . even in the great herbal of Leonhart Fuchs in the sixteenth century." Philosophers dealing with natural objects were more inclined to formal classification than medical botanists, who "were driven to classify [plants] simply because some sort of order was necessary for convenience in dealing with a large number of kinds . . . [h]ere and there . . . a slight feeling for [botanical] kinship emerges." What Arber is indicating is the problem of rote memorization of a large quantity of information. What seems possible, and entirely speculative, is that natural philosophers might have been engaged in creating or renovating the conceptual structures themselves—systems of classification—that would help order and "house" memory *specifically about physical plants*, whereas medical botanists might have been using ready-made memory structures of any kind to house the figures recorded in their herbals. In other words, a system of classification for most herbals might be thin or invisible in the books precisely because there were already existing, well-known structures and orders in which to "place" the wealth of figures representing knowledge and uses of plants, structures not specific to botanical knowledge, and not necessarily indicative of morphological or physically qualitative relationships pertaining only among plants.
25. Simon Schama, *Landscape and Memory*, 536–37. See also Andrew Cunningham, "The Culture of Gardens."
26. Schama, *Landscape and Memory*, 536.
27. Walter Benjamin, "On Some Motifs in Baudelaire," 185.
28. Ibid., 186.
29. Describing the aura of pressed plants to students in a visit to a colleague's classroom, a couple of students were curious if I had thought about photographing this aura, the way human "auras" are visualized by some photographic techniques, and seen unaided by some people. Even if an impression were to appear on whatever film a photographer would use for such an errand, by itself it would be meaningless. The aura Benjamin describes is a physical relationship that generates meaning; it is not a phenomenon captured by representation.
30. Benjamin, "The Work of Art in the Age of Mechanical Reproduction," 220–25.
31. Ibid., 221, 223.
32. Susan Stewart, *On Longing*, 150, 165.

33. Ibid., 157.
34. Celeste Olalquiaga, *The Artificial Kingdom*, 221, 222.
35. Ibid., 226–27. That these phrases remain in Latin in Olalquiaga's text underscores her treatment of them as distant phenomena, when they may not necessarily be so distant.
36. Ibid., 221–22.
37. Ibid., 222.
38. Giordano Bruno, *The Expulsion of the Triumphant Beast*, 235.
39. Though he too dismisses this order of the world as a very limited knowledge, Michel Foucault describes the outlines of the *episteme* that produced it in *The Order of Things*, in his chapter "The Prose of the World," 17–45.
40. Roger Chartier, *The Order of Books*, 80. Gesner was evidently deeply interested in classification in general. Not suprisingly for his education and his era, he was interested in botany, and compiled an herbarium; had he published what he found, he might have been given credit for naming some species attributed to later botanists. He also offered "the first formulation of the idea that genera should be denoted by substantive names. He was probably the earliest botanist who clearly expounded the distinction between a genus and a species." Arber, *Herbals*, 110–13, 167–68.
41. Yates, *The Art of Memory*, 79–80. The figures she does not recognize in the fresco suggest an additional quality of lists of virtues, vices, saints, forms of knowledge, and so on. Unfamiliar or locally important figures gain stature in proximity to known and widely shared ones, and reveal the particular orientation of a "list," not by what commonplaces it repeats, but by new or unusual additions in an otherwise familiar order. Stephen Parrish, teaching the (British) romantic poets at Cornell, would ask, "Who are the four greatest writers in the English language?" Students dutifully performed their memory of the trinity of this canon: Chaucer, Shakespeare, Milton—total agreement so far (whether sincere or otherwise is beside the point, we knew this was an exercise), and then the squabbling began over who the "last one" might be. Never mind that this exercise is a machine for exclusion. Parrish's point was, whomever the fourth writer turned out to be, that person signaled the preferences and outlook of the person completing the list—Byron? Yeats? A prose writer? A woman? An American? And the list became something else—a thought, an argument. With only one place to fill, of course we scrambled for an epitome of something, a way to complete this kind of allegorical list, condensing in that one name exactly the measure of significance in an imagined history acquired by the first three, either wrenching the canon wide open, closing it like a coffin lid (depending on one's view of canons), or more mildly signifying individual "taste," say, for a century or a style. The artificiality of the exercise was not intended to actually identify the "four greatest writers in the English language," but to demonstrate what kinds of questions were at play in choosing writers for study and comparison (and pleasure), as repositories of memory—whose writing do we feel obliged to remem-

ber and why? Whatever we chose (individually) gained its stature by juxtaposition with the first three, and changed their character in turn as well. To have a tantrum and shred the entire idea of a list would have avoided the central question: what is important to remember? Parrish also reminded us, with the slam of a heavy book on the table and an exclamation, that Byron "should be fun."

42. Umberto Eco was well aware of the memory arts and stock medieval divisions of knowledge and the cosmos when he designed the labyrinth of the library (and the plot itself) in *The Name of the Rose*. The book is thick with figures of memory and displays of the memory arts, not least of which is the layout of the Aedificium itself, according to an elaborate mystical pattern of numbers and significant divisions. Adso, struggling to decode the signs over the doors in the maze with William, remembers the last room, not by the textual fragment of the Apocalypse over the door, but by "a vision of a white horse" (171). This was why the technique was recommended—images are more memorable than words. The library, with its mysteries and legible codes repeated in its own geography, "was at once the celestial Jerusalem and an underground world on the border between terra incognita and Hades" (184). It was, of course, intended to be comprehensive. It included good as well as evil books, and the very question: how do you know? A dispute over one book plunges the whole structure into fire. It included good as well as evil people, a variety of lusts, weaknesses, and powers, each figured (hardly definitively) by appearance and gesture as well as behavior—again, how do you know? Long open loops of reading and meaning happen here anyway. Humorless Jorge was "the library's memory" (130), blind, bitter, poisoned by laughter—Eco's last word on structures of knowledge (including those embodied in individual people) whose hermetic closed-mindedness doom them to oblivion.
43. Yates, *The Art of Memory*, 31.
44. Antoine Du Verdier quoted in Chartier, *The Order of Books*, 87.
45. Olalquiaga, *The Artificial Kingdom*, 236.
46. Stewart, *On Longing*, 135.
47. Benjamin, "Unpacking My Library," 59–63.
48. Ibid., 66–67.
49. Benjamin, "One-Way Street," 60. This is Benjamin's epigraph, figuring the nature of his ramble with a proper noun—a person—as a "place" to begin, a memory of relationship which had its own transformative route.
50. Ibid., 66.
51. Gaston Bachelard, *The Poetics of Space*, 6, 14, 55, 241.
52. Ibid., 11, 15, 159.
53. Schama, *Landscape and Memory*, 577.
54. Ibid., 578.
55. David Abram, *The Spell of the Sensuous*, 176.
56. Stewart, *On Longing*, xiii.

My close friend and Virgo sister, Lisa Fischman, heard me out on the subject of necessary fragments for this album, suggested the possibility of a structured list, and said, "You're doing this because you're a Virgo." Another close friend, Rachel Buff, wove the astrological cosmos into ordinary (including intellectual) life when we were neighbors in Minneapolis. Both of these women were companions in graduate school, and astrology remains both a formal and informal framework for thinking through, interpreting, and experiencing things. We certainly learned more "legitimate" ones, but I doubt the legitimacy of ignoring any framework that can be made useful as well as satisfying, especially one that has a place already, somewhere, in actual living. We know, presumably, that what we learn to do in school does not describe or touch every aspect of life, but more than that, how we learn to think in school is not the only (or always the most valuable) way to think. That I would "do this because I am a Virgo"—and "as a Virgo" fuss over it in a note—is no less meaningful than whatever arrangement I might arrive at by another taxonomy or method. Virgo is a name for a certain way of being, a certain voice. A great deal of scholarship is written in the voice of Capricorn; some, especially luminous, is written as Pisces. The four elements structured my first book when I could not think of an intellectually and aesthetically satisfying way to divide up a potentially trackless field of information. Using twelve signs rather than four elements, this album has the advantage of acknowledging the fragmentary nature of the evidence while sketching a number of important things: a wide range of Ruth Nelson's knowledge, experience, and feeling over a transformative lifetime; the fact that a subject—me—is interested in what these fragments might say together this way; and that any subject has a heavy hand in arranging evidence of any kind. It is useful regardless of how much or little you know about astrological signs. Much: I'd encourage a contemporary reappropriation of interpretive tools beyond New Age sensibilities that can be at least as subtle and precise as the interpretive systems we use most, and welcome refinement. Little: I'd invite a willing suspension of disbelief and engagement with a pattern very much like a puzzle.

Needless to say, there both is and is not a bibliography for this sort of thing. The way I've understood astrology is dependent on practices of people I know, and the systematic training and practice of at least one professional, Pat Kaluza, who read my chart in Minneapolis in 1994. Relevant reference texts include Stephen Arroyo, *Astrology, Karma & Transformation: The Inner Dimensions of the Birth Chart*, second edition (Sebastopol, Calif.: CRCS Publications, 1992); Steven Forrest, *The Inner Sky: The Dynamic New Astrology for Everyone* (San Diego: ACS Publications, 1988); and Zipporah Dobyns, *Expanding Astrology's Universe* (San Diego: ACS Publications, 1988). These astrologers emphasize the interpretive capacity of astrological symbols, and describe their work as essentially discursive: astrology is a living language for reading and writing narratives, not a tool for prediction, and not "occult." They are influenced by psychological models (typically Jung's), by cultural anthropology, and comparative theology. Many astrological texts are understandably primers in basic reading and writing skills (and some are dismally reductive—Rachel Buff describes these as "cookbook astrology"); others explore more complex compositional issues regarding the symbolic work, for example, of particular planets. Though ambitious astrological practice draws on a long, complex history of allegorical interpretation, that history itself is unfortunately sketchy in these texts, but deserves sustained attention somewhere.

For this album, I used only the seasons of the calendar marked by the twelve signs of the zodiac; a life and its internal and external relationships over time are more fully mapped through planets and asteroids, and the twelve houses of the sky. These relationships are best understood as narrative fragments, articulated in a specific symbolic language. Each "location"—and combinations of locations (a planet in a sign in a house), as well as relationships between locations—is allegorical. In that respect the system is no different from conventional and contemporary psychoanalysis, except the "places" in the psychoanalytic cosmos are quite few—mother/father, the partitions of the self, and beyond that various forms of engulfment and annihilation. The astrological cosmos anticipates many worlds outside the self and family in relationship with an intrapsychic and spiritual life.

Feeling and thinking through this album discloses aspects of experience, not just in Ruth Nelson's life, but simultaneously in the act of interpreting, not easily arrived at by other means, certainly not within a brief exposition. The real work of astrology is narrative, and dialogic, not merely descriptive. It takes place in a relationship of interpretation and reflection—either between the astrologer and a subject, or within the subject herself, altogether within a larger matrix of relationships in the world (including the natural world).

One of the questions posed here is about the character of reflection in relation to the character of external order and discipline—Cancer and Capricorn, the worlds of home and career, summer and winter, mothers and fathers, informal and formal education, unconscious and conscious life, none of these immutable historical "facts" chained to one another, but flexible queries understood to have something to say to one another. It does matter that Ruth Nelson's life is the occasion: intellectual, emotional, and social relationships with and through natural things and places are the central concern of this book, and it was with that concern in mind that I started sorting, understanding with great frustration how little of her can be heard directly. But the album can't disclose "Ruth Nelson," and it wasn't intended to. Many "narratives," "reveries," and questions generally are possible here, involving an active reader, and that was the point.

Whether items actually fit the locations they've been placed in this album is one of the questions I imply, and the whole arrangement is a question of meaning that assumes, among other things, a dynamic relationship between public and private selves. What juxtapositions of available information make most sense? Are most satisfying (intuitively, aesthetically . . .)? For what purposes? In what ways are these usually public purposes tied to private habits of being? What kind of sense is most meaningful to you as you read books? What does that in turn say about how meaning happens in your own life? Is that different from how you seek out meaning, or purpose, in your work? Is there a reason there should be a difference?

1. Williams, *Aven Nelson of Wyoming*, 257.
2. Jane Ramsey, personal communication, July 2002.
3. Janet Robertson, *The Magnificent Mountain Women*, 119.
4. Ibid., 116–17.
5. Henry Pedersen Jr., *Those Castles of Wood*, 85–91. The building was razed in 1931. One wonders, had Wright's design been used, if it would have remained standing, though Wright had little acclaim when the building was commissioned.
6. Williams, *Aven Nelson of Wyoming*, 257; Robertson, *The Magnificent Mountain Women*, 117.
7. Ruth Nelson, travel notes, ANP, box 16, folder 9. Italicized passages are taken directly from Ruth Nelson's papers, whose collections are abbreviated s follows: ANP: Aven and Ruth Ashton Nelson Papers, American Heritage Center, Accession #400013; and RMH: collecting books housed at the Rocky Mountain Herbarium.
8. Williams, *Aven Nelson of Wyoming*, 257.
9. Ibid., 264.
10. Anna Maude Lute to Ruth Ashton Nelson, December 1931, ANP, box 16, folder 11, correspondence.
11. James Feucht to Ruth Ashton Nelson, 1 June 1961, ANP, box 16, folder 11, correspondence.
12. Jane Ramsey, personal communication, July 2002.
13. Robertson, *The Magnificent Mountain Women*, 116–17.
14. Ibid.
15. Ibid.
16. Beatrice Willard, "Foreword," Ruth Ashton Nelson, *Plants of Zion National Park*.
17. See Mary Arakelian, *Doc Orra Phelps*. Orra Phelps was almost an exact contemporary of Ruth Nelson's, and they narrowly missed crossing paths at Mount Holyoke—Orra transferred as a junior and graduated in 1918. Her mother had been a Mount Holyoke graduate, too. Intellectually curious, industrious, well educated, and unfortunately married to a man who was not capable of continuously providing for a large family—they had seven children—Mrs. Phelps (also named Orra) had a number of breakdowns, the first searing and severe. Afterwards, on the advice of her female physician, Mrs. Phelps made a promise to herself to take one day a week to do exactly as she pleased, leaving the house (and the chores and the children) to walk and botanize. Her interests and training had prepared her to be a professional, and she amassed respectable collections; her daughter deferred to her expertise for years as the "real" botanist.

    Orra, the daughter, learned several things from that breakdown. One was that her mother could in fact, and would again, break down, leaving Orra and her brother, Lawrence, to hold the household together in ways their parents were incapable. She also learned that botany was a rewarding thing to do, to share with her otherwise difficult mother, and to pursue on her own away from a chaotic household. Orra was a doctor for a living—she earned a medical degree from Johns Hopkins University—but botany was a serious avocation, part of her work as a naturalist and her interest in the Adirondack region where she became a well-known organizer of the Adirondack Mountain Club and guidebook writer. She eventually climbed all forty-six peaks over 4,000 feet, counting it among her loves and accomplishments. She mentored many young women through the Girl Scouts and her own informal teaching in the mountains. And, like Ruth, her "wilderness" was experienced most often with companions. She did not marry, a fact Arakelian (as well as introducer and historian of science Nancy Slack, who knew Orra) attributes to Mrs. Phelps's permanent demand for her daughter's availability in a family where the two eldest children ended up "parenting" the whole household and sacrificed some portion of their autonomous lives.

18. Ruth Nelson, travel notes, 15 July 1959, ANP, box 16, folder 9.
19. Williams, *Aven Nelson of Wyoming*, 257; Robertson, *The Magnificent Mountain Women*, 117.
20. Robertson, *The Magnificent Mountain Women*, 120.
21. Ibid., 120–21.
22. Jane Ramsey, personal communication, July 2002.
23. Williams, *Aven Nelson of Wyoming*, 317–23.
24. Ruth Nelson, manuscripts, 1961, ANP, box 13, folder 5.
25. Ruth Nelson, travel notes, ANP, box 16, folder 9. The alpine scene she describes on this spring day includes primroses, forget-me-nots, phlox, avens, bluebells, buttercups, marsh marigolds, kittentails, pennycress, and snowball saxifrage.
26. Ruth Nelson, travel notes, 21 May 1959, ANP, box 16, folder 9.
27. Robertson, *The Magnificent Mountain Women*, 118.
28. Ibid., 119.
29. Jane Ramsey, interview with Ruth Nelson, available in the Rocky Mountain National Park library; Robertson, *The Magnificent Mountain Women*, 117.
30. Tom Blaue quoted in Robertson, *The Magnificent Mountain Women*, 120.
31. Ruth Nelson, travel notes, ANP, box 16, folder 9. This scene includes species of forget-me-nots, clover, rock jasmine, and pennycress.
32. Williams, *Aven Nelson of Wyoming*, 272.
33. Tom Blaue quoted in Robertson, *The Magnificent Mountain Women*, 119.
34. Ruth Nelson, notes, ANP, box 16, folder 9.
35. Ruth and Aven Nelson, collecting book, RMH. As Ruth's full species designations indicate, these goldenrods were identified first by Gotthilf Mühlenberg and Carl Linnaeus, respectively.
36. Roger Williams, "Preface," Ruth Ashton Nelson, *Handbook of Rocky Mountain Plants* (1992), n.p.
37. Nelson, *Handbook of Rocky Mountain Plants*, 87–88.
38. Ibid., 208.

39. Ibid., 293.
40. Ibid., 342.
41. Ibid., 349.
42. Ruth Nelson, collecting book, RMH. This scene includes sunflower, wallflower, moss campion, Colorado and alpine columbine, spring beauty, penstemon, ragwort, bluebell, daisy, and buttercup, as well as Colorado bristlecone pine. Ruth refers to her collecting notes here as if they were elsewhere, but this scene is in fact sketched in the collecting book, framed by collection numbers (#6474 and #6475, from 29 and 30 June on her way to Loveland, Colorado). The species she names here are numbered in the list several pages later. Evidently she filled a mostly empty page after continuing to list for a couple days.
43. Ruth Nelson, travel notes, ANP, box 16 folder 9.

### HABEAS CORPUS

1. Ruth Nelson, travel journal, ANP, Accession #400013, box 16, folder 9.
2. Tom Blaue quoted in Janet Robertson, *The Magnificent Mountain Women*, 120.
3. Regarding the taxonomy of married female botanists, see Nancy Slack, "Nineteenth-Century American Women Botanists" and "Botanical and Ecological Couples: A Continuum of Relationships."
4. Table of contents, four of the five groups of chapters in Helena Pycior, Nancy Slack, and Pnina Abir-Am, eds., *Creative Couples in the Sciences*.
5. "Comparative Study of Couples along Disciplinary and Transdisciplinary Lines" is the fifth and last group of chapters in *Creative Couples in the Sciences*.
6. Judith Jordan, "Empathy and Self Boundaries," 69.
7. Judith Jordan, "The Meaning of Mutuality," 89.
8. James Elkins, *The Object Stares Back*, 140.
9. Ibid., 156.
10. Gregory Bateson, *Mind and Nature*, 7–9.
11. Holmes, *The Young John Muir*, 266.
12. Ibid., 268.
13. Ibid., 270, 271.
14. Ibid., 273, 274.
15. Holmes, *The Young John Muir*, 3.
16. Ibid., 14–15.
17. Ibid., 16, 248.
18. Ibid.
19. Terrence O'Connor, "Therapy for a Dying Planet," 154.
20. Scott Slovic, "Robert Michael Pyle: A Portrait," in Robert Michael Pyle, *Walking the High Ridge*, 137.
21. My mother's experience of their marriage and household was certainly different from mine, and the same is true for my daughter in relation to my mother. I grew up with my grandparents to some extent, but not at all in the way my mother did. My grandparents were always "old" to me; my mother is a "young" grand-

mother to my daughter, vibrantly active, and her household feels different to my daughter than it did to me, partly of course because all the principal players have different lives than they did when I lived there. What is still striking across four or five generations is the fund of variability and partial echoes from one generation to the next, compelling separations as well as similarities. I strongly suspect intellectual lineages work the same way, and moreover that these lineages form in some communication with patterns of repetition and difference learned with people outside the circle of study and work. In what ways do we learn how to read and study from people other than teachers, in activities we don't think of as "reading" and "studying"? When a book or an idea is compelling enough to work with, what patterns of interest and recognition are we invoking and reinventing? We see the results of this process as a body of intellectual work—facts, ideas, interpretations—but how is its form related to the form of learning beyond official study, and how do learned intellectual habits in turn reverberate, for better and for worse, in relationships outside work?

COLLECTING

I used Roger Williams's *Aven Nelson of Wyoming* (1984) for the sequence of events that brought Aven down the mountain to Alice as she was dying, for the text of letters Aven sent to his family and Ruth, as well as for the fact that Aven went southwest looking for *Calhounia.* I put these events in the context of Aven's field notes and apparent collecting habits. In the field notes it's clear that Aven left *Calhounia* aside as soon as he came home, and that emotional or recreational errands and botanical ones coincided regularly throughout his life. Williams provides a complete list of Aven Nelson's publications, technical and otherwise.

All my information about actual plants Aven and Ruth collected and the geography of their collecting comes from their field notebooks, which are housed at the University of Wyoming Rocky Mountain Herbarium. They are not catalogued formally, but the dated, numbered series is continuous, except for the striking break in numbering after Aven's marriage to Ruth.

Aven Nelson's professional addresses, manuscripts, lecture notes from Harvard and the University of Wyoming, and correspondence with a curious public, as well as his own copy of his *New Manual of Botany of the Central Rocky Mountains (Vascular Plants)* (1909), are available at the University of Wyoming American Heritage Center, Laramie, Wyoming, in two collections: Department of Botany Records (accession #545001), and Aven and Ruth Ashton Nelson Papers (accession #400013). These collections, as well as Williams's biography and the records of the Department of Geology and Geophysics (accession #545004), are the sources for information about Nelson's involvement at the University of Wyoming's science camp and his enjoyment of the place.

I quoted from Janet Robertson's account of Ruth in her book, *The Magnificent Mountain Women: Adventures in the Colorado Rockies* (1990), which was based on interviews with Ruth and her friends before Ruth's death in 1987. Manuscripts of

Ruth's popular articles (for the magazine *Green Thumb*) are at the American Heritage Center, Aven and Ruth Ashton Nelson Papers, Accession #400013, box 13. From the same collection, I quoted from Ruth's "Report on the Study of the Plants of McKinley National Park," 1945, box 12, folder 4.

Ranger naturalists in Denali gave informative talks on park history, ethnobotany, plant and animal ecology, and environmental issues the week I was there, all of them women: Jessica Brillhart, Jen McWeeny, Lori Rome, and Sheila Isanaka.

The text of this essay in italics is drawn from my journals, including the summer journal of specific trips in 2001, where I listed plants I found and identified (or tried to). I also collected plants—on forest service lands and in the Red Desert, as well as near Laramie and in Fairbanks—to learn how to press and dry them, and how to keep track of what they were and where they grew during the long process of collecting and later mounting them. Spending hours with just a few plants on Rogers Canyon Road drove home the necessity of collecting them to study later.

# Bibliography


PRIMARY SOURCES

Aven and Ruth Ashton Nelson Papers. University of Wyoming American Heritage Center. Accession #400013.

Aven and Ruth Ashton Nelson collecting books. University of Wyoming Rocky Mountain Herbarium.

Frederic and Edith Clements Papers. University of Wyoming American Heritage Center. Accession #1678.

University of Wyoming Department of Botany Records. University of Wyoming American Heritage Center. Accession #545001.

University of Wyoming Department of Geology and Geophysics Records. University of Wyoming American Heritage Center. Accession #545004.

SECONDARY SOURCES

Abram, David. *The Spell of the Sensuous*. New York: Vintage Books, 1997.

Al-shehbaz, Ihsan A. "Reed Clark Rollins December 7, 1911–April 28, 1998." http://www.nap.edu/readingroom/books/biomems/rrollins.html, downloaded 26 December 2001.

Anderson, Benedict. *Imagined Communities: Reflections on the Origin and Spread of Nationalism*. London: Verso, 1983.

Arakelian, Mary. *Doc Orra Phelps, M.D.: Adirondack Naturalist and Mountaineer*. Introduction by Nancy Slack. Ithaca: North Country Books, 2000.
Arber, Agnes. *Herbals, Their Origin and Evolution, a Chapter in the History of Botany, 1470–1670*. 2nd ed. Cambridge: Cambridge University Press, 1953/1912.
Arrien, Angeles. *The Tarot Handbook: Practical Applications of Ancient Visual Symbols*. Sonoma: Arcus, 1987.
Ashworth, William Jr. "Emblematic Natural History of the Renaissance," p. 17–37. In *Cultures of Natural History*, edited by N. Jardine, J. A. Secord, and E. C. Spary. Cambridge: Cambridge University Press, 1996.
Bachelard, Gaston. *The Poetics of Space*. Trans. Maria Jolas. Boston: Beacon Press, 1969.
Bailey, Liberty Hyde. *The Nature Study Idea*. New York: Doubleday, 1903.
Bailey, William Whitman. *The Botanical Collector's Handbook*. Salem, Mass.: George A. Bates, 1881.
Basso, Keith. *Wisdom Sits in Places: Landscape and Language among the Western Apache*. Albuquerque: University of New Mexico Press, 1996.
Bateson, Gregory. *Mind and Nature: A Necessary Unity*. Toronto: Bantam Books, 1979.
Bateson, Mary Catherine. *Composing a Life*. New York: Atlantic Monthly Press, 1989.
Benjamin, Walter. "On Some Motifs in Baudelaire," p. 155–200. In *Illuminations*, edited by Hannah Arendt. New York: Schocken Books, 1968.
———. "One-Way Street," p. 61–94. In *Reflections*, edited by Peter Demetz. New York: Schocken Books, 1978.
———. "Unpacking My Library," p. 59–67. In *Illuminations*, edited by Hannah Arendt. New York: Schocken Books, 1968.
———. "The Work of Art in the Age of Mechanical Reproduction," p. 217–51. In *Illuminations*, edited by Hannah Arendt. New York: Schocken Books, 1968.
Billington, Ray Allen. *Frederick Jackson Turner: Historian, Scholar, Teacher*. New York: Oxford University Press, 1973.
Bonta, Marcia Myers. *Women in the Field: America's Pioneering Women Naturalists*. College Station: Texas A&M University Press, 1991.
Bridson, David, Donald Wendel, and James White. *Printmaking in the Service of Botany: 21 April to 31 July 1986, Catalogue of an Exposition*. Pittsburgh: Hunt Institute for Botanical Documentation, 1986.
Brown, William E. *Denali: Symbol of the Alaskan Wild*. Denali National Park: Alaska Natural History Association, 1993.
Bruno, Giordano. *The Expulsion of the Triumphant Beast*. Translated and edited by Arthur Imerti. New Brunswick, N.J.: Rutgers University Press, 1964.
Buchanan, Handasyde. *Nature into Art: A Treasury of Great Natural History Books*. New York: Mayflower Books, 1979.
Chartier, Roger. *The Order of Books*. Trans. Lydia Cochrane. Stanford, Calif.: Stanford University Press, 1994.

Clough, Wilson. *A History of the University of Wyoming 1887–1964*. Laramie: University of Wyoming, 1965.
Comstock, Anna. *The Comstocks of Cornell*. New York: Comstock, 1953.
Coulter, John M. *Manual of the Botany (Phaenogamia and Pteridophyta) of the Rocky Mountain Region, from New Mexico to the British Boundary*. New York: Ivison, Blakeman, Taylor, 1885.
Cunningham, Andrew. "The Culture of Gardens," p. 38–56. In *Cultures of Natural History*, edited by N. Jardine, J. A. Secord, and E. C. Spary. Cambridge: Cambridge University Press, 1996.
Derrida, Jacques. *Archive Fever: A Freudian Impression*. Trans. Eric Prenowitz. Chicago: University of Chicago Press, 1996.
Dickenson, Victoria. *Drawn From Life: Science and Art in the Portrayal of the New World*. Toronto: University of Toronto Press, 1998.
Dorn, Harold. *The Geography of Science*. Baltimore: Johns Hopkins University Press, 1991.
Dorn, Robert. *Vascular Plants of Wyoming*. 2nd ed. Cheyenne, Wyo.: Mountain West, 1992.
Duncan, Kate, and Larry Gay. *Hiking and Biking (and Fishing) in the Medicine Bow National Forest*. Laramie: Larry Gay, 1996.
Dupree, A. Hunter. *Asa Gray, 1810–1880*. Cambridge: Harvard University Press, 1959.
Eco, Umberto. *The Name of the Rose*. San Diego: Harcourt Brace, 1984.
Elkins, James. *The Object Stares Back: On the Nature of Seeing*. San Diego: Harcourt Brace, 1996.
Ewan, Joseph. *Rocky Mountain Naturalists*. Denver: University of Denver Press, 1950.
Foucault, Michel. *The Order of Things: An Archaeology of the Human Sciences*. New York: Vintage Books, 1994.
Fryxell, Fritiof. *The Tetons: Interpretations of a Mountain Landscape*. Moose, Wyo.: Grand Teton Natural History Association, 1995/1938.
Ganong, William F. *The Teaching Botanist: A Manual of Information Upon Botanical Instruction*. New York: Macmillan, 1899.
George, Wilma. *Animals and Maps*. Berkeley: University of California Press, 1969.
Gray, Asa. *Introduction to Structural and Systematic Botany and Vegetable Physiology*. New York: Ivison, Blakeman, Taylor, 1878.
———. *Lessons in Botany and Vegetable Physiology*. New York: Ivison, Blakeman, Taylor, 1879/1857.
———. *Manual of Botany*. Edited by Merritt Lyndon Fernald. 8th ed. New York: D. Van Nostrand, 1970.
Holmes, Steven. *The Young John Muir: An Environmental Biography*. Madison: University of Wisconsin Press, 1999.
Hooker, William J. "On the Botany of America." *American Journal of Science* 9(1825): 263–84.

Hultén, Eric. *Flora of Alaska and Neighboring Territories: A Manual of the Vascular Plants.* Stanford, Calif.: Stanford University Press, 1968.

Johnson, Derek, Linda Kershaw, Andy MacKinnon, and Jim Pojar. *Plants of the Western Boreal Forest and Aspen Parkland.* Edmonton, Alberta: Lone Pine, 1995.

Jordan, Judith. "Empathy and Self Boundaries," p. 67–80. In *Women's Growth in Connection: Writings from the Stone Center,* edited by Judith Jordan, Alexandra Kaplan, Jean Baker Miller, Irene Stiver, and Janet Surrey. New York: Guilford Press, 1991.

———. "The Meaning of Mutuality," p. 81–96. In *Women's Growth in Connection: Writings from the Stone Center,* edited by Judith Jordan, Alexandra Kaplan, Jean Baker Miller, Irene Stiver, and Janet Surrey. New York: Guilford Press, 1991.

Keeney, Elizabeth. *The Botanizers: Amateur Scientists in Nineteenth-Century America.* Chapel Hill: University of North Carolina Press, 1992.

Klein, Richard, and Deana Klein. *Research Methods in Plant Science.* New York: American Museum of Natural History, 1970.

Kuklick, Henrika, and Robert Kohler, eds. *Science in the Field.* Chicago: University of Chicago Press, 1996.

Luria, Aleksandr. *The Mind of a Mnemonist: A Little Book about a Vast Memory.* Trans. Lynn Soloratoff. New York: Basic Books, 1968.

Magoc, Chris. *Yellowstone: The Creation and Selling of an American Landscape, 1870–1903.* Albuquerque: University of New Mexico Press, 1999.

Maguire, Bassett. "Highlights of Botanical Exploration in the New World," p. 209–46. In *Fifty Years of Botany: Golden Jubilee Volume of the Botanical Society of America,* edited by William Campbell Steere. New York: McGraw-Hill, 1958.

Marshall, Mark. *Yellowstone Trails: A Hiking Guide.* Yellowstone National Park: Yellowstone Association, 1999.

Muir, John. *The Yellowstone Park.* Silverthorn, Colo.: Vistabooks, 1999/1898.

Nelson, Aven. *New Manual of Botany of the Central Rocky Mountains (Vascular Plants).* New York: American Book, 1909.

———. *The Red Desert of Wyoming and Its Forage Resources.* Washington, D.C.: Government Printing Office, 1898.

Nelson, Ruth Ashton. *Handbook of Rocky Mountain Plants.* Tucson: Dale Stuart King, 1969.

———. *Handbook of Rocky Mountain Plants.* Revised and edited by Roger Williams. Toronto: Key Porter Books, 1992.

———. *Mountain Wild Flowers of Colorado and Adjacent Areas.* With Rhoda N. Roberts. Denver: Denver Museum of Natural History, 1967.

———. *Plants of Rocky Mountain National Park.* Illustrations by Beatrice Willard. Washington, D.C.: Government Printing Office, 1933. Revised editions 1953 by GPO and Estes Park, Colo.: Rocky Mountain Nature Association, 1976, 1982.

———. *Plants of Zion National Park: Wildflowers, Trees, Shrubs and Ferns.* Foreword by Beatrice Willard, illustrations by Tom Blaue. Springdale, Utah: Zion Natural History Association, 1976.

———. *Wild Flowers of Wyoming*. With drawings by Guinevere Kennington. Laramie: Wyoming Agricultural Extension Service, 1936. Rev. eds. 1962, 1968; revised by Ronald L. Hartman, 1984.

O'Connor, Terrence. "Therapy for a Dying Planet," p. 149–55. In *Ecopsychology: Restoring the Earth, Healing the Mind*, edited by Theodore Roszak, Mary E. Gomes, and Allen D. Kanner. San Francisco: Sierra Club Books, 1995.

Olalquiaga, Celeste. *The Artificial Kingdom: A Treasury of the Kitsch Experience*. New York: Pantheon Books, 1998.

Peattie, Daniel Culross. "On the Popularization of Botany," p. 456–66. In *Fifty Years of Botany: Golden Jubilee Volume of the Botanical Society of America*, edited by William Campbell Steere. New York: McGraw-Hill, 1958.

Pedersen, Henry Jr. *Those Castles of Wood: The Story of Early Lodges of Rocky Mountain National Park and Pioneer Days of Estes Park, Colorado*. Estes Park: Henry Pedersen Jr., 1993.

Pratt, Verna E., and Frank G. Pratt. *Wildflowers of Denali National Park and Interior Alaska*. Anchorage: Alaskakrafts, 1993.

Pycior, Helena, Nancy Slack, and Pnina Abir-Am, eds. *Creative Couples in the Sciences*. New Brunswick, N.J.: Rutgers University Press, 1996.

Pyle, Robert Michael. *The Thunder Tree: Lessons from an Urban Wildland*. New York: Lyons Press, 1998/1993.

———. *Walking the High Ridge: Life as Field Trip*. With "Robert Michael Pyle: A Portrait," by Scott Slovic. Minneapolis: Milkweed Editions, 2000.

———. *Wintergreen: Rambles in a Ravaged Land*. New York: Charles Scribner's Sons, 1986.

Reckling, Frederick W., and JoAnn B. Reckling. *Samuel Howell "Doc" Knight: Mr. Wyoming University*. Laramie: University of Wyoming Alumni Association, 1998.

Richards, Joan. *Angles of Reflection: Logic and a Mother's Love*. W. H. Freeman, 2000.

Richards, Mary Bradshaw. *Camping Out in the Yellowstone, 1882*. Edited by William Slaughter. Salt Lake City: University of Utah Press, 1994/1910.

Robertson, Janet. *The Magnificent Mountain Women: Adventures in the Colorado Rockies*. Lincoln: University of Nebraska Press, 1990.

Rollins, Reed C. "Taxonomy of Higher Plants," p. 192–208. In *Fifty Years of Botany: Golden Jubilee Volume of the Botanical Society of America*, edited by William Campbell Steere. New York: McGraw-Hill, 1958.

Rossiter, Margaret. *Women Scientists in America: Struggles and Strategies to 1940*. Baltimore: Johns Hopkins University Press, 1982.

Schama, Simon. *Landscape and Memory*. New York: Vintage Books, 1996/1995.

Schneider, Bill. *Hiking Grand Teton National Park*. Guilford, Conn.: Globe Pequot Press, 1998.

Shaw, Richard J. *Plants of Yellowstone and Grand Teton National Parks*. Camano Island, Wash.: Wheelwright, n.d.
———. *Vascular Plants of Grand Teton National Park & Teton County: An Annotated Checklist*. Moose, Wyo.: Grand Teton Natural History Association, 1992.
———. *Wildflowers of Grand Teton and Yellowstone National Parks*. Rev. ed. Camano Island, Wash.: Wheelwright, 1992.
Slack, Nancy. "Botanical and Ecological Couples: A Continuum of Relationships," p. 235–53. In *Creative Couples in the Sciences*, edited by Helena Pycior, Nancy Slack, and Pnina Abir-Am. New Brunswick, N.J.: Rutgers University Press, 1996.
———. "Nineteenth-Century American Women Botanists: Wives, Widows, and Work," p. 77–103. In *Uneasy Careers and Intimate Lives*, edited by Pnina Abir-Am and D. Outram. New Brunswick, N.J.: Rutgers University Press, 1987.
Smith, Diane. *Letters from Yellowstone*. New York: Penguin, 2000.
Spellenberg, Richard. *The Audubon Society Field Guide to North American Wild Flowers, Western Region*. New York: Knopf, 1979.
Stearn, William T. Introduction to *Species Plantarum*, by Carl Linnaeus. Facsimile of first 1753 edition. 2 vols. London: Ray Society, 1957.
Stevens, William Chase. *Introduction to Botany*. Boston: D.C. Heath, 1902.
Stewart, Susan. *On Longing: Narratives of the Miniature, the Gigantic, the Souvenir, the Collection*. Baltimore: Johns Hopkins University Press, 1984.
Steyermark, Cora. *Behind the Scenes*. St. Louis: Missouri Botanical Garden, 1984.
Tompkins, Jane. *A Life in School: What the Teacher Learned*. Reading, Mass.: Addison-Wesley, 1996.
Tuan, Yi-fu. *Topophilia: A Study of Environmental Perception, Attitudes, and Values*. Englewood Cliffs, N.J.: Prentice-Hall, 1974.
Weber, William A. *Rocky Mountain Flora*. Niwot, Colo.: University Press of Colorado, 1976.
White, Richard. "'Are You an Environmentalist or Do You Work for a Living?': Work and Nature," p. 171–85. In *Uncommon Ground: Toward Reinventing Nature*, edited by William Cronon. New York: Norton, 1995.
———. *The Organic Machine: The Remaking of the Columbia River*. New York: Hill and Wang, 1995.
Whitson, Tom, ed. *Weeds of the West*. Jackson, Wyo.: Western Society of Weed Science, 1991.
Williams, Roger. *Aven Nelson of Wyoming*. Boulder: Colorado Associated University Press, 1984.
Winter, David, and Andrea DiNoto. *The Pressed Plant: The Art of Botanical Specimens, Nature Prints, and Sun Pictures*. New York: Stewart, Tabori & Chang, 1999.
Yates, Frances. *The Art of Memory*. London: Routledge and Kegan Paul, 1966.

AMERICAN LAND AND LIFE SERIES

Bachelor Bess: The Homesteading Letters of Elizabeth Corey, 1909–1919
Edited by Philip L. Gerber

Botanical Companions: A Memoir of Plants and Place
By Frieda Knobloch

Circling Back: Chronicle of a Texas River Valley
By Joe C. Truett

Edge Effects: Notes from an Oregon Forest
By Chris Anderson

Exploring the Beloved Country: Geographic Forays into American Society and Culture
By Wilbur Zelinsky

Father Nature: Fathers as Guides to the Natural World
Edited by Paul S. Piper and Stan Tag

The Follinglo Dog Book: From Milla to Chip the Third
By Peder Gustav Tjernagel

Great Lakes Lumber on the Great Plains: The Laird, Norton Lumber Company in South Dakota
By John N. Vogel

Hard Places: Reading the Landscape of America's Historic Mining Districts
By Richard V. Francaviglia

Landscape with Figures: Scenes of Nature and Culture in New England
By Kent C. Ryden

Living in the Depot: The Two-Story Railroad Station
By H. Roger Grant

Main Street Revisited: Time, Space, and Image Building in Small-Town America
By Richard V. Francaviglia

Mapping American Culture
Edited by Wayne Franklin and Michael C. Steiner

Mapping the Invisible Landscape: Folklore, Writing, and the Sense of Place
By Kent C. Ryden

Mountains of Memory: A Fire Lookout's Life in the River of No Return Wilderness
By Don Scheese

The People's Forests
By Robert Marshall

Pilots' Directions: The Transcontinental Airway and Its History
Edited by William M. Leary

Places of Quiet Beauty: Parks, Preserves, and Environmentalism
By Rebecca Conard

Reflecting a Prairie Town: A Year in Peterson
Text and photographs by Drake Hokanson

A Rural Carpenter's World: The Craft in a Nineteenth-Century New York Township
By Wayne Franklin

Salt Lantern: Traces of an American Family
By William Towner Morgan

Signs in America's Auto Age: Signatures of Landscape and Place
By John A. Jakle and Keith A. Sculle

Thoreau's Sense of Place: Essays in American Environmental Writing
Edited by Richard J. Schneider